S0-AKJ-249

Longman Mathematical Texts

Special functions of mathematical physics and chemistry

Third edition

Ian N. Sneddon

Simson Professor of Mathematics,
The University of Glasgow

Longman
London and New York

Longman Group Limited London

Associated companies, branches and representatives throughout the world

Published in the United States of America by Longman Inc., New York

First published by Oliver and Boyd 1956
Second edition 1961
Reprinted 1966
Third edition by Longman Group Ltd 1980

British Library Cataloguing in Publication Data

Sneddon, Ian Naismith
Special functions of mathematical physics and chemistry. – 3rd ed. – (Longman mathematical texts).
1. Functions, Special 2. Mathematical Physics
3. Chemistry – Mathematics
I. Title
515′.5′02453 QC20.7.F/ 79-40969

ISBN 0-582-44396-2

Filmset in Northern Ireland at The Universities Press (Belfast) Ltd.
Printed in England by McCorquodale (Newton) Ltd., Newton-le-Willows, Lancashire.

Preface to the third edition

In the main this edition is a reprint of the second edition – with some fresh material and additional problems included. The major part of the new material consists of a discussion of some general features of linear ordinary differential equations of the second order, and of the method of finding a solution of such an equation in the form of a contour integral. The inclusion of this topic is made possible by the fact that in the twenty-five years since this little book first appeared courses in mathematics for scientists and engineers have included much more function theory than they did. However, it is my hope that the book will continue to be of value to users of mathematics unacquainted with the basics of the theory of functions of a complex variable.

IAN N. SNEDDON

Glasgow *April 1979*

Contents

1
Introduction

§1. The origin of special functions

The special functions of mathematical physics arise in the solution of partial differential equations governing the behaviour of certain physical quantities. Probably the most frequently occurring equation of this type in all physics is Laplace's equation

$$\nabla^2\psi = 0 \tag{1.1}$$

satisfied by a certain function ψ describing the physical situation under discussion. The mathematical problem consists of finding those functions which satisfy equation (1.1) and also satisfy certain prescribed conditions on the surfaces bounding the region being considered. For example, if ψ denotes the electrostatic potential of a system, ψ will be constant over any conducting surface. The shape of these boundaries often makes it desirable to work in curvilinear coordinates q_1, q_2, q_3 instead of in rectangular Cartesian coordinates x, y, z. In this case we have relations

$$x = x(q_1, q_2, q_3), \quad y = y(q_1, q_2, q_3), \quad z = z(q_1, q_2, q_3) \tag{1.2}$$

expressing the Cartesian coordinates in terms of the curvilinear coordinates. If equations (1.2) are such that

$$\frac{\partial x}{\partial q_i}\frac{\partial x}{\partial q_j} + \frac{\partial y}{\partial q_i}\frac{\partial y}{\partial q_j} + \frac{\partial z}{\partial q_i}\frac{\partial z}{\partial q_j} = 0$$

when $i \neq j$ we say that the coordinates q_1, q_2, q_3 are **orthogonal curvilinear coordinates.** The element of length dl is then given by

$$dl^2 = h_1^2\, dq_1^2 + h_2^2\, dq_2^2 + h_3^2\, dq_3^2 \tag{1.3}$$

where

$$h_i^2 = \left(\frac{\partial x}{\partial q_i}\right)^2 + \left(\frac{\partial y}{\partial q_i}\right)^2 + \left(\frac{\partial z}{\partial q_i}\right)^2 \tag{1.4}$$

and it can easily be shown that

$$\nabla^2\psi = \frac{1}{h_1h_2h_3}\left\{\frac{\partial}{\partial q_1}\left(\frac{h_2h_3}{h_1}\frac{\partial\psi}{\partial q_1}\right)+\frac{\partial}{\partial q_2}\left(\frac{h_3h_1}{h_2}\frac{\partial\psi}{\partial q_2}\right)+\frac{\partial}{\partial q_3}\left(\frac{h_1h_2}{h_3}\frac{\partial\psi}{\partial q_3}\right)\right\}. \tag{1.5}$$

One method of solving Laplace's equation consists of finding solutions of the type

$$\psi = Q_1(q_1)Q_2(q_2)Q_3(q_3)$$

by substituting from (1.5) into (1.1). We then find that

$$\frac{1}{Q_1}\frac{\partial}{\partial q_1}\left(\frac{h_2h_3}{h_1}\frac{\partial Q_1}{\partial q_1}\right)+\frac{1}{Q_2}\frac{\partial}{\partial q_2}\left(\frac{h_3h_1}{h_2}\frac{\partial Q_2}{\partial q_2}\right)+\frac{1}{Q_3}\frac{\partial}{\partial q_3}\left(\frac{h_1h_2}{h_3}\frac{\partial Q_3}{\partial q_3}\right)=0.$$

If, further, it so happens that

$$\frac{h_2h_3}{h_1} = f_1(q_1)F_1(q_2, q_3)$$

etc., then this last equation reduces to the form

$$F_1(q_2, q_3)\frac{1}{Q_1}\frac{\mathrm{d}}{\mathrm{d}q_1}\left\{f_1(q_1)\frac{\mathrm{d}Q_1}{\mathrm{d}q_1}\right\}+F_2(q_3, q_1)\frac{1}{Q_2}\frac{\mathrm{d}}{\mathrm{d}q_2}\left\{f_2(q_2)\frac{\mathrm{d}Q_2}{\mathrm{d}q_2}\right\}$$
$$+F_3(q_1, q_2)\frac{1}{Q_3}\frac{\mathrm{d}}{\mathrm{d}q_3}\left\{f_3(q_3)\frac{\mathrm{d}Q_3}{\mathrm{d}q_3}\right\}=0.$$

Now, in certain circumstances, it is possible to find three functions $g_1(q_1)$, $g_2(q_2)$, $g_3(q_3)$ with the property that

$$F_1(q_2, q_3)g_1(q_1)+F_2(q_3, q_1)g_2(q_2)+F_3(q_1, q_2)g_3(q_3)\equiv 0.$$

When this is so, it follows immediately that the solution of Laplace's equation (1.1) reduces to the solution of three self-adjoint ordinary linear differential equations

$$\frac{\mathrm{d}}{\mathrm{d}q_i}\left\{f_i\frac{\mathrm{d}Q_i}{\mathrm{d}q_i}\right\}-g_iQ_i=0, \qquad i=1, 2, 3. \tag{1.6}$$

It is the study of differential equations of this kind which leads to the special functions of mathematical physics. The adjective 'special' is used in this connection because here we are not, as in analysis, concerned with the general properties of functions, but only with the properties of functions which arise in the solution of special problems.

To take a particular case, consider the cylindrical polar coordinates (ρ, ϕ, z) defined by the equations

$$x = \rho \cos \phi, \quad y = \rho \sin \phi, \quad z = z$$

for which $h_1 = 1$, $h_2 = \rho$, $h_3 = 1$. From equation (1.5) we see that, for these coordinates, Laplace's equation is of the form

$$\frac{\partial^2 \psi}{\partial \rho^2} + \frac{1}{\rho}\frac{\partial \psi}{\partial \rho} + \frac{1}{\rho^2}\frac{\partial^2 \psi}{\partial \phi^2} + \frac{\partial^2 \psi}{\partial z^2} = 0. \tag{1.7}$$

If we now make the substitution

$$\psi = R(\rho)\Phi(\phi)Z(z), \tag{1.8}$$

we find that equation (1.7) may be written in the form

$$\frac{1}{R}\left(\frac{\mathrm{d}^2 R}{\mathrm{d}\rho^2} + \frac{1}{\rho}\frac{\mathrm{d}R}{\mathrm{d}\rho}\right) + \frac{1}{\rho^2 \Phi}\frac{\mathrm{d}^2 \Phi}{\mathrm{d}\phi^2} + \frac{1}{Z}\frac{\mathrm{d}^2 Z}{\mathrm{d}z^2} = 0.$$

This shows that if Φ, Z, R satisfy the equations

$$\frac{\mathrm{d}^2 \Phi}{\mathrm{d}\phi^2} + n^2 \Phi = 0. \tag{1.9a}$$

$$\frac{\mathrm{d}^2 Z}{\mathrm{d}z^2} - \xi^2 Z = 0. \tag{1.9b}$$

$$\frac{\mathrm{d}^2 R}{\mathrm{d}\rho^2} + \frac{1}{\rho}\frac{\mathrm{d}R}{\mathrm{d}\rho} + \left(\xi^2 - \frac{n^2}{\rho^2}\right)R = 0 \tag{1.9c}$$

respectively, then the function (1.8) is a solution of Laplace's equation (1.7). The study of these ordinary differential equations will lead us to the special functions appropriate to this coordinate system. For instance, equation (1.9a) may be taken as the equation *defining* the circular functions. In this context $\sin(n\phi)$ is defined as that solution of (1.9a) which has value 0 when $\phi = 0$ and $\cos(n\phi)$ as that which has value 1 when $\phi = 0$ and the properties of the functions derived therefrom, cf. Problem 1.4 below. Similarly, equation (1.9b) defines the exponential functions. In actual practice we do not proceed in this way merely because we have already encountered these functions in another context and from their familiar properties studied their relation to equations (1.9a) and (1.9b). The situation with respect to equation (1.9c) is different; we cannot express its solution in terms of the elementary functions of analysis, as we were able to do with the other two equations. In this case we define new functions in terms of the solutions of this equation and by investigating the series solutions of the equations derive the properties of the functions so defined. If we change the independent variable from ρ to $x = \xi\rho$ and write $R(\rho) = y(\xi\rho)$ we see that $y(x)$ must satisfy the equation

$$\frac{d^2y}{dx^2} + \frac{1}{x}\frac{dy}{dy} + \left(1 - \frac{n^2}{x^2}\right)y = 0. \tag{1.10}$$

Equation (1.10) is called **Bessel's equation** and solutions of it are called **Bessel functions.** Bessel functions are of great importance in theoretical physics; they are discussed in Chapter 4.

§2. Solution in series of linear differential equations

We shall have occasion to discuss ordinary linear differential equations of the second order with variable coefficients whose solutions cannot be obtained in terms of the elementary functions of mathematical analysis. In such cases one of the standard procedures is to derive a pair of linearly independent solutions in the form of infinite series and from these series to compute tables of standard solutions. With the aid of such tables the solution appropriate to any given initial conditions may then be readily found. The object of this note is to outline briefly the procedure to be followed in these instances; for proofs of the theorems quoted the reader is referred to the standard textbooks – see, e.g., Burkill (1975), Ch. III.

2.1 Ordinary points of a linear differential equation

A function is called **analytic** at a point if it is possible to expand it in a Taylor series valid in some neighbourhood of the point. This is equivalent to saying that the function is single-valued and possess derivatives of all orders at the point in question. In the equations we shall consider the coefficients will be analytic functions of the independent variable except possibly at certain isolated points.

An **ordinary** point $x = a$ of the second order differential equation

$$y'' + \alpha(x)y' + \beta(x)y = 0 \tag{2.1}$$

is one at which the coefficients α, β are analytic functions. It can be shown that *at any ordinary point every solution of the equation is analytic.* Furthermore *if the Taylor expansions of $\alpha(x)$ and $\beta(x)$ are valid in the range $|x-a|<R$ the Taylor expansion of the solution is valid for the same range.* As a consequence, if $\alpha(x)$ and $\beta(x)$ are polynomials in x the series solution of (2.1) is valid for *all* values of x.

When, as is usually the case, $\alpha(x)$ and $\beta(x)$ are polynomials of low degree, the solution is most easily found by assuming a power series of the form

$$y=\sum_{r=0}^{\infty} c_r(x-a)^r \tag{2.2}$$

for the solution and determining the coefficients $c_0, c_1, c_2, \ldots,$ by direct substitution of (2.2) into (2.1) and equating coefficients of successive powers of x to zero.

The simplest equation of this type is

$$y''+y=0. \tag{2.3}$$

Substituting a solution of the type (2.2) with $a=0$ into this equation we find that, if the equation is to be satisfied,

$$\sum_{r=0}^{\infty} r(r-1)c_r x^{r-2}+\sum_{r=0}^{\infty} c_r x^r=0.$$

The series on the left is equivalent to

$$\sum_{r=0}^{\infty} (r+1)(r+2)c_{r+2}x^r$$

so that, equating coefficients of x^r, we see that the equation is satisfied by a solution of type (2.2) provided that the coefficients are connected by the relation

$$(r+1)(r+2)c_{r+2}+c_r=0. \tag{2.4}$$

The coefficients c_0, c_1 are determined by the prescribed values of y, y' at $x=0$, and the others are determined by equation (2.4). From this relation it follows that the solution is

$$y=c_0\left(1-\frac{x^2}{2!}+\frac{x^4}{4!}-\ldots\right)+c_1\left(x-\frac{x^3}{3!}+\frac{x^5}{5!}-\ldots\right). \tag{2.5}$$

An equation of the kind (2.4) which determines the subsequent coefficients in terms of the first two is called a **recurrence relation.**

2.2 Regular singular points

If either of the functions $\alpha(x)$, $\beta(x)$ is not analytic at the point $x=a$, we say that this point is a **singular** point of the differential equation. When the functions $\alpha(x)$, $\beta(x)$ are of such a nature that the differential equation may be written in the form

$$(x-a)^2y''+(x-a)p(x)y'+q(x)y=0 \tag{2.6}$$

where $p(x)$ and $q(x)$ are analytic at the point $x=a$ we say that this point is a **regular singular** point of the differential equation.

If $x=a$ is a regular singular point of the equation (2.6) it can be shown that there exists at least one solution of the form

$$y=\sum_{r=0}^{\infty} c_r(x-a)^{\rho+r} \tag{2.7}$$

which is valid in some neighbourhood of $x=a$. More specifically, if the Taylor expansions for $p(x)$, $q(x)$ are valid for $|x-a|<R$, the solution (2.7) is valid in the same range.

Putting

$$p(x)=\sum_{r=0}^{\infty} p_r(x-a)^r, \quad q(x)=\sum_{r=0}^{\infty} q_r(x-a)^r \tag{2.8}$$

and substituting the expansions (2.7) and (2.8) into (2.6) we see that for the equation (2.6) to be satisfied we must have

$$\sum_{r=0}^{\infty} c_r(\rho+r)(\rho+r-1)(x-a)^{\rho+r} + \sum_{s=0}^{\infty} p_s(x-a)^s \sum_{r=0}^{\infty} c_r(\rho+r)(x-a)^{\rho+r} + \sum_{s=0}^{\infty} q_s(x-a)^s \sum_{r=0}^{\infty} c_r(x-a)^{\rho+r}=0. \tag{2.9}$$

Equating to zero the coefficient of $(x-a)^\rho$ we have the relation

$$c_0\rho(\rho-1)+p_0c_0\rho+q_0c_0=0$$

so that if $c_0\neq 0$ we have the quadratic equation

$$\rho^2+(p_0-1)\rho+q_0=0 \tag{2.10}$$

for the determination of ρ. This is known as the **indicial equation.** Similarly if we equate to zero the coefficient of $(x-a)^{\rho+r}$ we obtain the relation

$$c_r(\rho+r)(\rho+r-1)+\sum_{s=0}^{r}\{p_s(\rho+r-s)+q_s\}c_{r-s}=0$$

which may be written in the form

$$c_r\{(\rho+r)(\rho+r-1)+p_0(\rho+r)+q_0\} + \sum_{s=1}^{r}\{p_s(\rho+r-s)+q_s\}c_{r-s}=0. \quad (2.11)$$

Equation (2.11) gives the two possible values ρ_1, ρ_2 of ρ. If we take one of these values, ρ_1 say, and substitute it in the **recurrence relation** (2.11) we obtain the corresponding value of the coefficients c_r and hence the solution

$$y_1(x)=\sum_{r=0}^{\infty} c_r(x-a)^{r+\rho_1}.$$

In a similar way the root ρ_2 of the indicial equation leads to the solution

$$y_2(x)=\sum_{r=0}^{\infty} c_r'(x-a)^{r+\rho_2}.$$

Three distinct cases arise according to the nature of the roots of the indicial equation.

Case (i) $\rho_1-\rho_2$ *neither zero nor an integer*
In these circumstances the solutions $y_1(x)$ and $y_2(x)$ are linearly independent and the general solution of equation (2.6) is of the form

$$y=\sum_{r=0}^{\infty} c_r(x-a)^{r+\rho_1}+\sum_{r=0}^{\infty} c_r'(x-a)^{r+\rho_2}. \quad (2.12)$$

Case (ii) $\rho_1=\rho_2$
If $\rho_1=\rho_2$ the solutions $y_1(x)$ and $y_2(x)$ are identical (except, possibly, for a multiplicative constant). The general solution of the equation can be shown to be $c_1y_1(x)+c_2y_2(x)$ where

$$\left.\begin{aligned} y_1(x)&=(x-a)^{\rho_1}\sum_{r=0}^{\infty} c_r(x-a)^r,\\ y_2(x)&=y_1(x)\log(x-a)+(x-a)^{\rho_1}\sum_{r=0}^{\infty}\left(\frac{\partial c_r}{\partial\rho}\right)_{\rho=\rho_1}(x-a)^r. \end{aligned}\right\} \quad (2.13)$$

Case (iii) $\rho_2 = \rho_1 - n$ *where n is a positive integer*
In this case all the coefficients in one of the solutions from some point onwards are either infinite or indeterminate. It can be shown that the appropriate solutions are

$$\left.\begin{aligned} y_1(x) &= (x-a)^{\rho_1} \sum_{r=0}^{\infty} c_r (x-a)^r, \\ y_2(x) &= g_n y_1(x) \log(x-a) + (x-a)^{\rho_2} \sum_{r=0}^{\infty} b_r (x-a)^r, \end{aligned}\right\} \tag{2.14}$$

where g_n is the coefficient of x^n in the expansion of

$$\frac{x^{n+1}}{\{y_1(x+a)\}^2} \exp\left[-\int_0^x up(u)\, du\right].$$

It may happen that $g_n = 0$ in which case $y_2(x)$ does not contain a logarithmic term.

2.3 The point at infinity

In many problems we wish to find solutions of differential equations of the type (2.1) which are valid for large values of x. We seek solutions in the form of infinite series with variable $1/x$. If we make the transformation

$$x = \frac{1}{\xi}$$

the 'point at infinity' is taken into the origin on the ξ-axis. With this change of variable equation (2.1) becomes

$$\frac{d^2y}{d\xi^2} + \left\{\frac{2}{\xi} - \frac{1}{\xi^2}\alpha\left(\frac{1}{\xi}\right)\right\}\frac{dy}{d\xi} + \frac{1}{\xi^4}\beta\left(\frac{1}{\xi}\right)y = 0. \tag{2.15}$$

§3. Solution in form of a contour integral

It is also possible to derive solutions of linear differential equations in the form of contour integrals. Again we shall confine our attention to equations of the second order of the form

$$p(z)\frac{d^2w}{dz^2} + q(z)\frac{dw}{dz} + r(z)w = 0, \tag{3.1}$$

whose coefficients p, q, r will usually be polynomials of low degree. In many instances we shall be particularly interested in **self-adjoint** equations.

$$\frac{\mathrm{d}}{\mathrm{d}z}\left[p(z)\frac{\mathrm{d}w}{\mathrm{d}z}\right]+r(z)w=0, \tag{3.2}$$

which is of the form (3.1) with $q=p'$.

3.1 The Fourier–Laplace method

In this method we assume that equation (3.1) has a solution of the form

$$w(z)=\int_C e^{isz}f(s)\,\mathrm{d}s, \tag{3.3}$$

in which the function f and the contour C have to be determined. For this form

$$\frac{\mathrm{d}w}{\mathrm{d}z}=i\int_C se^{isz}f(s)\,\mathrm{d}s,$$

so that if we have a term of the form $zw'(z)$ in equation (3.1) it will yield a term

$$iz\int_C se^{isz}f(s)\,\mathrm{d}s,$$

which we may transform (using the rule for integration by parts) to

$$\Delta_C[sf(s)e^{isz}]-\int_C e^{isz}\frac{\mathrm{d}}{\mathrm{d}s}\{sf(s)\}\,\mathrm{d}s,$$

where $\Delta_C[\phi(s)]$ denotes the change in value of the function ϕ as we go from one end of the contour C in the s-plane to the other. In a similar way, since

$$\frac{\mathrm{d}^2w}{\mathrm{d}z^2}=-\int_C s^2e^{isz}f(s)\,\mathrm{d}s,$$

we can show that

$$z\frac{\mathrm{d}^2w}{\mathrm{d}z^2}=\Delta_C[is^2f(s)e^{isz}]-i\int_C e^{isz}\frac{\mathrm{d}}{\mathrm{d}s}\{s^2f(s)\}\,\mathrm{d}s$$

and that

$$z^2\frac{d^2w}{dz^2}=\Delta_C\left[\left\{izs^2f(s)-\frac{d}{ds}[s^2f(s)]\right\}e^{isz}\right]$$
$$+\int_C e^{isz}\frac{d^2}{ds^2}\{s^2f(s)\}\,ds.$$

Proceeding in this way and substituting into equation (3.1) we transform that equation to the form

$$\Delta_C[F(z,s)]+\int_C e^{isz}\mathbb{L}f(s)\,ds=0,$$

where $\mathbb{L}$ is a differential operator whose order is that of the highest degree among the polynomials p, q and r. In order, therefore, for the contour integral (3.3) to be a solution of the equation (3.1) we must take f to be a solution of the differential equation

$$\mathbb{L}f=0 \tag{3.4}$$

and the contour C to be such that

$$\Delta_C[F(z,s)]=0. \tag{3.5}$$

The differential equation $\mathbb{L}f=0$ may be no easier to solve than the original equation (3.1) but it should be noticed that we do not require to find the general solution of this equation; for our immediate purposes a particular solution will do.

To illustrate the method we consider the differential equation

$$z\frac{d^2w}{dz^2}+(2\nu+1)\frac{dw}{dz}+zw(z)=0. \tag{3.6}$$

Making use of the above results we find that in this case

$$\mathbb{L}f(s)=-i\left[(s^2-1)\frac{df}{ds}-(2\nu-1)sf(s)\right]$$

and

$$F(z,s)=(s^2-1)f(s)e^{isz}.$$

Writing $\mathbb{L}f=0$ in the form

$$\frac{f'(s)}{f(s)}=(2\nu-1)\frac{s}{s^2-1}$$

we see that equation (3.6) has a solution of the form

$$w(z)=\int_C e^{isz}(s^2-1)^{\nu-\frac{1}{2}}\,ds, \tag{3.7}$$

where the contour C is arbitrary except that it must satisfy the condition

$$\Delta_C[(s^2-1)^{\nu+\frac{1}{2}}e^{isz}]=0.$$

It is easily verified that this condition is satisfied if C is a figure of eight which encircles the point $s=+1$ in a counter-clockwise direction and the point $s=-1$ in a clockwise direction.

In some cases it is more appropriate to assume a solution of the form

$$w(z)=\int_C e^{pz}g(p)\,dp.$$

The same method then reduces the problem of finding g and C to that of finding a particular solution of a differential equation and devising a contour to satisfy a condition of the type (3.5).

To illustrate the method we consider the solution of the equation

$$z\frac{d^2w}{dz^2}+(\gamma-z)\frac{dw}{dz}-\alpha w=0.$$

Since

$$\frac{dw}{dz}=\int_C e^{pz}pg(p)\,dp,$$

$$z\frac{dw}{dz}=\Delta_C[e^{pz}pg(p)]-\int_C e^{pz}\frac{d}{dp}\{pg(p)\}\,dp,$$

$$z\frac{d^2w}{dz^2}=\Delta_C[e^{pz}p^2g(p)]-\int_C e^{pz}\frac{d}{dp}\{p^2g(p)\}\,dp,$$

it follows that g must be a solution of the equation

$$-\frac{d}{dp}\{p^2g(p)\}+\gamma pg(p)+\frac{d}{dp}\{pg(p)\}-\alpha g(p)=0.$$

Writing this equation in the form

$$\frac{g'(p)}{g(p)}=\frac{\alpha-1}{p}+\frac{\gamma-\alpha-1}{p-1},$$

we see that we may take

$$g(p)=p^{\alpha-1}(p-1)^{\gamma-\alpha-1}.$$

The condition on C is

$$\Delta_C[p(p-1)g(p)e^{pz}]=0,$$

which is equivalent to

$$\Delta_C[p^{\alpha}(p-1)^{\gamma-\alpha}e^{pz}]=0.$$

3.2 The Euler method

A second method is associated with the name of Euler.

To illustrate the method we consider the equation

$$Q(z)\frac{d^2w}{dz^2}+L(z)\frac{dw}{dz}+\lambda w=0, \tag{3.8}$$

in which $Q(z)$ is a **quadratic** function of z, $L(z)$ a **linear** function and λ a **constant.** We now choose a quadratic function p_2, a linear function p_1 and a constant μ such that

$$\begin{aligned} Q(z)&=p_2(z),\\ L(z)&=-(\mu-1)p_2'(z)-p_1(z),\\ \lambda&=\tfrac{1}{2}\mu(\mu-1)p_2''(z)+\mu p_1'(z); \end{aligned}$$

in other words we take

$$\begin{aligned} p_2(z)&=Q(z),\\ p_1(z)&=-L(z)-(\mu-1)Q'(z), \end{aligned} \tag{3.9}$$

where μ is a root of the quadratic equation

$$\tfrac{1}{2}Q''(0)\mu(\mu-1)+\mu L'(0)+\lambda=0. \tag{3.10}$$

We now seek solutions of equation (3.8) of the form

$$w(z)=\int_C (s-z)^{\mu}f(s)\,ds, \tag{3.11}$$

in which the function f in the integrand and the line C are, as yet, unspecified. Since

$$\frac{dw}{dz}=-\mu\int_C (s-z)^{\mu-1}f(s)\,ds$$

and

$$\frac{d^2w}{dz^2} = \mu(\mu-1)\int_C (s-z)^{\mu-2}f(s)\,ds,$$

we see that if (3.11) is a solution of (3.8) then f and C must be chosen to be such that

$$\int_C [\{p_2(z)+p_2'(z)(s-z)+\tfrac{1}{2}p_2''(z)(s-z)^2\}\mu(\mu-1)(s-z)^{\mu-2}f(s)$$
$$+\{p_1(z)+p_1'(z)(s-z)\}\mu(s-z)^{\mu-1}f(s)]\,ds=0. \quad (3.12)$$

Using the results

$$p_2(z)+p_2'(z)(s-z)+\tfrac{1}{2}p_2''(z)(s-z)^2 = p_2(s),$$
$$p_1(z)+p_1'(z)(s-z) = p_1(s),$$

and the rule for integrating by parts, we see that equation (3.12) is equivalent to the differential equation

$$\frac{d}{ds}\{p_2(s)f(s)\}-p_1(s)f(s)=A,$$

where A is an arbitrary constant, for the determination of the function f and that the condition for the contour C is

$$\Delta_C F(z,s)=0$$

with

$$F(z,s)=\mu(s-z)^{\mu-1}p_2(s)f(s)-A(s-z)^{\mu}.$$

Since we are seeking any solution and **not** the general solution of the differential equation for $f(s)$ we may take $A=0$. Hence we take f to be any solution the equation

$$\frac{f'(s)}{f(s)}+\frac{L(s)+\mu Q'(s)}{Q(s)}=0 \quad (3.13)$$

and C to be any contour such that

$$\Delta_C(s-z)^{\mu-1}Q(s)f(s)=0. \quad (3.14)$$

In the case in which

$$Q(z)=(z-z_1)(z-z_2)$$

we may take

$$f(s)=(s-z_1)^{-k_1}(s-z_2)^{-k_2},$$

where

$$k_1 = \mu - \frac{L(z_1)}{z_2 - z_1}, \qquad k_2 = \mu + \frac{L(z_2)}{z_2 - z_1} \tag{3.15}$$

with μ a root of the equation

$$\mu^2 + \{L'(0) - 1\}\mu + \lambda = 0. \tag{3.16}$$

Hence

$$w(z) = \int_C (s - z_1)^{-k_1}(s - z_2)^{-k_2}(s - z)^{\mu}\, ds \tag{3.17}$$

is a solution of the equation

$$(z - z_1)(z - z_2)\frac{d^2w}{dz^2} + L(z)\frac{dw}{dz} + \lambda w = 0, \tag{3.18}$$

where k_1, k_2 are given by equations (3.15) with μ any root of the quadratic equation (3.16); the contour C must satisfy the condition

$$\Delta_C (s - z)^{\mu-1}(s - z_1)^{1-k_1}(s - z_2)^{1-k_2} = 0. \tag{3.19}$$

If equation (3.8) is self-adjoint $L(z) = Q'(z)$, so that we may take

$$f(s) = \{Q(s)\}^{-\mu-1}.$$

In particular the self-adjoint equation

$$\frac{d}{dz}\left\{(z - z_1)(z - z_2)\frac{dw}{dz}\right\} + \lambda w(z) = 0$$

has solution

$$\int_C (s - z_1)^{-\mu-1}(s - z_2)^{-\mu-1}(s - z)^{\mu}\, ds, \tag{3.20}$$

where the contour C must be such that

$$\Delta_C[(s - z)^{\mu-1}(s - z_1)^{-\mu}(s - z_2)^{-\mu}] = 0 \tag{3.21}$$

In this case we may take for C:

(*a*) a simple closed curve enclosing the points $s = z_1$, $s = z$ but not the point $s = z_2$;

(*b*) a simple closed curve enclosing the points $s = z_2$, $s = z$ but not the point $s = z_1$;

(*c*) a figure-of-eight going once round $s=z_1$ in the counter-clockwise direction and once round $s=z_2$ in the clockwise direction.

For example, for Legendre's equation

$$(1-z^2)\frac{d^2w}{dz^2}-2z\frac{dw}{dz}+\nu(\nu+1)=0 \tag{3.22}$$

we may take $Q(z)=z^2-1$, $L(z)=2z$, $\lambda=-\nu(\nu+1)$. It then follows from equations (3.20), (3.21) and (3.16) that

$$w(z)=\int_C (s^2-1)^{-\mu-1}(s-z)^{\mu}\, ds \tag{3.23}$$

is a solution of (3.22) if $\mu=\nu$ or $-\nu-1$, and C is a contour of any one of the types (*a*)–(*c*) listed above.

Taking $\mu=-\nu-1$ we get the solution

$$w(z)=A\int_C \frac{(s^2-1)^{\nu}}{(s-z)^{\nu+1}}\, ds. \tag{3.24}$$

Hence if we take C to be a simple closed curve C_1 enclosing the points $s=z$ and $s=1$, we find that in the case when $\nu=n$, a positive integer

$$w(z)=A\int_{C_1} \frac{(s^2-1)^n}{(s-z)^{n+1}}\, ds$$

is a solution of Legendre's equation such that

$$w(1)=A\int_{C_1} \frac{(s+1)^n}{s-1}\, ds=2\pi i \,.\, 2^n \,.\, A.$$

Hence

$$w(z)=\frac{1}{2^{n+1}\pi i}\int_{C_1} \frac{(s^2-1)^n}{(s-z)^{n+1}}\, ds$$

is a solution of (3.22) with $\nu=n$ and $w(1)=1$. Now, by a well-known corollary of Cauchy's theorem, if $f(s)$ is holomorphic in the region of the s-plane bounded by C_1,

$$f^{(n)}(z)=\frac{n!}{2\pi i}\int_{C_1} \frac{f(s)\, ds}{(s-z)^{n+1}},$$

so that the above solution may be written in the form

$$w(z)=\frac{1}{2^n n!}\frac{d^n}{dz^n}(z^2-1)^n. \tag{3.25}$$

By taking a different choice for C in equation (3.24) we can obtain a different kind of solution of Legendre's equation (cf. Problems 1.14 and 1.15).

§4. Self-adjoint linear equations of the second order

We have seen already that self-adjoint equations

$$\frac{d}{dx}\left[p(x)\frac{dy}{dx}\right]+q(x)y=0 \tag{4.1}$$

play a special role in the theory of linear differential equations of the second order. For that reason we shall give (without proofs) an account of the principal properties of such equations. (For proofs, the reader is referred to Burkill (1975) Ch. III.

4.1 Separation and comparison theorems

Two solutions of (4.1) are said to be **linearly dependent** if there exists a non-zero constant c such that $y_1(x)=cy_2(x)$. If no such constant exists y_1 and y_2 are said to be **linearly independent.** It is easily shown that if we define the **Wronskian** of y_1 and y_2 by

$$W(y_1, y_2; x)=y_1(x)y_2'(x)-y_1'(x)y_2(x), \tag{4.2}$$

then a necessary and sufficient condition for y_1 and y_2 to be linearly independent is that $W(y_1, y_2; x)$ never vanishes.

Also if $p(x)>0$ for all values of x in interval J on the real line then we can easily derive **Abel's formula**

$$p(x)W(y_1, y_2; x)=k, \qquad \forall x\in J, \tag{4.3}$$

where k is a constant which is zero only if the solutions y_1 and y_2 of (4.1) are linearly dependent. From this result we deduce

Lemma 4.1. Two solutions y_1 and y_2 of the self-adjoint equation (4.1) have a common zero if and only if they are linearly independent.

From Abel's formula and this lemma we can immediately derive

Sturm's separation theorem. If y_1 and y_2 are linearly independent solutions of a self-adjoint linear equation of the second order, then between two consecutive zeros of y_1 there is precisely one zero of y_2

and from this theorem we can in turn prove

Sturm's comparison theorem. If a solution $y(x)$ of

$$\frac{d}{dx}\, p(x)\frac{dy}{dx}+q(x)y=0 \qquad p(x)>0, \tag{4.4}$$

has consecutive zeros at $x=x_1$ and $x=x_2$, $(x_1<x_2)$ and if $Q(x)>q(x)$ with strict inequality holding for at least one point of the interval $[x_1, x_2]$, a solution of

$$\frac{d}{dx}\, p(x)\frac{dy}{dx}+Q(x)y=0 \tag{4.5}$$

which vanishes at $x=x_1$ will vanish again at some point on the open interval (x_1, x_2).

Another useful result is

The theorem of numerical comparison. If $y(x)$ is a solution of

$$\frac{d}{dx}\, p(x)\frac{dy}{dx}+q(x)y=0, \qquad p(x)>0,$$

satisfying $y(a)=y_0$, $y'(a)=y_1$ and if $z(x)$ is a solution of

$$\frac{d}{dx}\, p(x)\frac{dz}{dx}+Q(x)z=0 \qquad Q(x)>q(x),$$

satisfying $z(a)=y_0$, $z'(a)=y_1$, then in any neighbourhood to the right of the point $x=a$ in which $y(x)$ and $z(x)$ never vanish (except possibly at $x=a$),

$$|y(x)|>|z(x)|.$$

The Sturm comparison theorem may be used to obtain upper and lower bounds for the distance between two successive zeros of a solution of a self-adjoint linear equation of the second order.

To facilitate the discussion we assume that the equation has been reduced to the **normal form**

$$\frac{d^2y}{dx^2}+P(x)y=0, \tag{4.6}$$

(cf. Problem 1.17). The principal result is a corollary of the Sturm comparison theorem; it may be stated as:

Lemma 4.2. If $P(x)$ is always positive and

$$m^2 \leqslant P(x) \leqslant M^2, \tag{4.7}$$

then if δ is the distance between any two successive zeros of any solution whatsoever of the equation (4.6),

$$\frac{\pi}{M} \leqslant \delta \leqslant \frac{\pi}{m}. \tag{4.8}$$

For an illustration of the application of this lemma the reader is referred to Problem 1.18.

4.2 Eigenvalues and eigenfunctions

The problem is to find non-trivial solutions of the two-point boundary value problem

$$\frac{d^2y}{dx^2}+\lambda y=0, \qquad 0 \leqslant x \leqslant a$$

$$y(0)=y(a)=0.$$

It is elementary to show that non-trivial solutions exist only if λ takes one of the values $\{\lambda_n\}_{n=1}^{\infty}$ where $\lambda_n = n^2\pi^2/a^2$. When $\lambda=\lambda_n$ the appropriate solution of the problem is $c_n\psi_n(x)$, where c_n is an arbitrary constant and

$$\psi_n(x)=(2/a)^{\frac{1}{2}} \sin(n\pi x/a).$$

It will be observed that this function has the property that

$$\int_0^a \psi_m(x)\psi_n(x)\,dx = \delta_{mn} \tag{4.9}$$

where δ_{mn} is the Kronecker delta defined by the relations

$$\delta_{mn} = \begin{cases} 1 & \text{if } m=n \\ 0 & \text{if } m \neq n. \end{cases}$$

Using the notation

$$\langle f, g\rangle = \int_0^a f(x)g(x)\,dx \tag{4.10}$$

we can rewrite equation (4.9) in the form

$$\langle \psi_m, \psi_n\rangle = \delta_{mn}. \tag{4.11}$$

A set of functions with this property is said to be **orthonormal** on $[0, a]$.

We say that the numbers λ_n are **eigenvalues** and that the functions ψ_n are **eigenfunctions** of the stated boundary value problem.

It will be observed that for this simple problem:

(*a*) there is a smallest eigenvalue $\lambda_1\ (=\pi^2/a^2)$;
(*b*) there is an infinite sequence $\{\lambda_n\}_{n=1}^{\infty}$ of eigenvalues, and $\lambda_n \to \infty$;
(*c*) the eigenfunctions can be normalised to form an orthonormal set.

In general, it can be shown that **there exists a sequence $\{\lambda_n\}_{n=1}^{\infty}$ of eigenvalues of the boundary value problem**

$$\frac{d}{dx}\left[p(x)\frac{dy}{dx}\right]+\{q(x)+\lambda\rho(x)\}y=0, \quad y(a)=y(b)=0, \quad \rho(x)>0,$$

with $\lambda_1<\lambda_2<\ldots<\lambda_n<\ldots$, $\lambda_n \to\infty$ as $n\to\infty$ and a corresponding sequence $\{\phi_n\}_{n=1}^{\infty}$ of eigenfunctions. In addition, $\phi_n(x)$ has precisely n zeros on $[a, b]$.

If ϕ_m, ϕ_n are eigenfunctions corresponding to the eigenvalues λ_m, λ_n, then

$$\frac{d}{dx}\{p(x)\phi_m'(x)\}+\{q(x)+\lambda_m\rho(x)\}\phi_m(x)=0,$$

$$\frac{d}{dx}\{p(x)\phi_n'(x)\}+\{q(x)+\lambda_n\rho(x)\}\phi_n(x)=0,$$

with $\phi_m(a)=\phi_n(a)=0$, $\phi_m(b)=\phi_m(a)=0$. Multiplying the first of these equations by ϕ_n, the second by ϕ_m and subtracting we obtain the relation

$$(\lambda_m-\lambda_n)\rho(x)\phi_m(x)\phi_n(x)=\frac{d}{dx}\{p(x)[\phi_m(x)\phi_n'(x)-\phi_m'(x)\phi_n(x)]\}.$$

Integrating over $[a, b]$ and using the boundary values of ϕ_m and

ϕ_n, we find that if $\lambda_m \neq \lambda_n$, i.e. $m \neq n$,

$$\langle \phi_m, \phi_n \rangle_\rho = 0 \quad (m \neq n)$$

where

$$\langle f, g \rangle_\rho = \int_a^b f(x)g(x)\rho(x)\,dx$$

If a separate calculation shows that

$$\langle \phi_m, \phi_m \rangle_\rho = c_m^2,$$

then the functions ψ_m defined by

$$\psi_m(x) = c_m^{-1}\phi_m(x) \tag{4.12}$$

satisfy the condition

$$\langle \psi_m, \psi_n \rangle_\rho = \delta_{mn}. \tag{4.13}$$

A sequence of functions $\{\psi_m\}_{m=1}^{\infty}$ satisfying the condition (4.13) is said to be **orthonormal in** $[a, b]$ **with respect to the weight function** $\rho(x)$. If $\{\psi_m\}_{m=1}^{\infty}$ is such a sequence it follows, of course, that the sequence $\{\Psi_m\}_{m=1}^{\infty}$ defined by the equation

$$\Psi_m(x) = \{\rho(x)\}^{\frac{1}{2}}\psi_m(x)$$

is orthonormal in the sense of equation (4.11).

A great deal has been written on the problem of expanding an arbitrary function defined on a finite interval $[a, b]$ in terms of an orthonormal set $\{\Psi_m\}_{m=1}^{\infty}$ on that interval. A proper treatment is given in first courses in functional analysis. Here, we shall proceed in a purely formal manner.

If we can write

$$f(x) = \sum_{m=1}^{\infty} c_m \Psi_m(x)$$

then

$$\langle f, \Psi_m \rangle = \sum_{j=1}^{\infty} c_j \langle \Psi_j, \Psi_m \rangle = c_m$$

so that we have the formal expansion

$$f(x) = \sum_{m=1}^{\infty} \langle f, \Psi_m \rangle \Psi_m(x). \tag{4.14}$$

The series on the right side of this equation is called the **Fourier**

series of f and the constants $\langle f, \Psi_m\rangle = c_m^{-1}\langle f, \rho^{\frac{1}{2}}\phi_m\rangle$ are called the **Fourier coefficients** of f.

From equation (4.14) we deduce immediately that

$$\langle f, g\rangle = \sum_{m=1}^{\infty} \langle f, \Psi_m\rangle\langle\Psi_m, g\rangle \tag{4.15}$$

and, in particular, if we write $\|f\|^2 = \langle f, f\rangle$

$$\|f\|^2 = \sum_{m=1}^{\infty} \{\langle f, \Psi_m\rangle\}^2. \tag{4.16}$$

This last equation is usually referred to as **Parseval's identity.**

Orthogonal functions arise in a natural way in approximation theory. One example occurs in least square approximation (cf. Problem 1.19). Another example is provided by Gaussian integration formulae; it can be shown that (§1.3 of Wendroff (1966))

$$\int_a^b w(x)f(x)\,dx = \sum_{k=1}^{n} w_k f(x_k) + R_n(f), \tag{4.17}$$

where $\{x_k\}_{k=1}^{\infty}$ are the zeros of polynomials p_n (of degree n) which are orthogonal on $[a, b]$ with weight function w,

$$w_k = \frac{1}{p_n'(x_k)} \int_a^b \frac{w(x)p_n(x)}{x - x_k}\,dx \tag{4.18}$$

and $R_n(f)$ is a remainder term, given by the formula

$$R_n(f) = Cf^{(2n)}(\xi), \qquad \xi \in (a, b), \tag{4.19}$$

with the constant C defined by

$$C = \frac{1}{(2n)!} \int_a^b [(x - x_1)\dots(x - x_n)]^2 w(x)\,dx. \tag{4.20}$$

§5. The gamma function and related functions

In developing series solutions of differential equations and in other formal calculations it is often convenient to make use of properties of gamma and beta functions. The integral

$$\Gamma(n) = \int_0^{\infty} e^{-x}x^{n-1}\,dx \tag{5.1}$$

converges if $n > 0$ and defines the **gamma function.** Similarly if

$m>0$, $n>0$ the **beta function** is defined by the equation

$$B(m, n)=\int_0^1 x^{m-1}(1-x)^{n-1}\,dx. \tag{5.2}$$

It is then easily shown that

(a) $\Gamma(1)=1$,
(b) $\Gamma(n+1)=n\Gamma(n)$,
(c) $\Gamma(n+1)=n!$ if n is a positive integer,
(d) $B(m, n)=2\int_0^{\frac{1}{2}\pi} \sin^{2m-1}\theta \cos^{2n-1}\theta\, d\theta$,
(e) $B(m, n)=\Gamma(m)\Gamma(n)/\Gamma(m+n)$,
(f) $\Gamma(\frac{1}{2})=\sqrt{\pi}$,
(g) $\Gamma(p)\Gamma(1-p)=\pi \operatorname{cosec}(p\pi), \quad 0<p<1$,
(h) $\Gamma(\frac{1}{2})\Gamma(2n)=2^{2n-1}\Gamma(n)\Gamma(n+\frac{1}{2})$ – the **duplication formula,**
(i) $\Gamma(z+1)=\lim_{n\to\infty} n!\, n^z/(z+1)(z+2)\ldots(z+n), \quad z>0.$

When n is a negative fraction $\Gamma(n)$ is defined by means of equation (b); for example

$$\Gamma(-\tfrac{3}{2})=\frac{\Gamma(-\frac{1}{2})}{-\frac{3}{2}}=\frac{\Gamma(\frac{1}{2})}{(-\frac{3}{2})(-\frac{1}{2})}=\frac{4\Gamma(\frac{1}{2})}{3}.$$

By means of the result (i) we can derive an interesting expression for **Euler's constant,** γ, which is defined by the equation

$$\gamma=\lim_{n\to\infty}\left(1+\frac{1}{2}+\ldots+\frac{1}{n}-\log n\right)=0.5772\ldots \tag{5.3}$$

From (i) we have

$$\frac{d}{dz}\{\log\Gamma(z+1)\}=\lim_{n\to\infty}\left(\log n-\frac{1}{z+1}-\frac{1}{z+2}-\ldots-\frac{1}{z+n}\right),$$

so that letting $z\to 0$ we obtain the result

$$\gamma=-\left[\frac{d}{dz}\log\Gamma(z+1)\right]_{z=0}, \tag{5.4}$$

and from (5.1) we find

$$\gamma=-\int_0^\infty e^{-t}\log t\,dt. \tag{5.5}$$

Integrating by parts we see that

$$\int_z^\infty e^{-t}\log t\,dt=+e^{-z}\log z+\int_z^\infty \frac{e^{-t}}{t}\,dt,$$

so that

$$-\gamma = \lim_{z\to\infty}\left(\int_z^\infty \frac{e^{-t}}{t}\,dt + \log z\right). \tag{5.6}$$

Regarded as a function of the complex variable z, $\Gamma(z)$ is single-valued and holomorphic over the whole complex plane, except for the points $z=-n$ $(n=0, 1, 2, \ldots)$ where it possesses simple poles with residue $(-1)^n n!$ Consequently, its reciprocal $1/\Gamma(z)$ is an entire function with simple zeros at the points $z=-n$ $(n=0, 1, 2, \ldots)$.

Closely related to the gamma function are the **exponential-integral** $\mathrm{ei}(x)$ defined by the equation

$$\mathrm{ei}(x) = \int_x^\infty \frac{e^{-u}}{u}\,du, \qquad x>0, \tag{5.7}$$

and the **logarithmic-integral** $\mathrm{li}(x)$ defined by

$$\mathrm{li}(x) = \int_0^x \frac{du}{\log u}, \tag{5.8}$$

which are themselves connected by the relation

$$\mathrm{ei}(x) = -\mathrm{li}(e^{-x}). \tag{5.9}$$

Other integrals of importance are the **sine and cosine integrals** $\mathrm{Ci}(x)$, $\mathrm{Si}(x)$, which are defined by the equations

$$\mathrm{Ci}(x) = -\int_x^\infty \frac{\cos u}{u}\,du, \qquad \mathrm{Si}(x) = \int_0^x \frac{\sin u}{u}\,du \tag{5.10}$$

and whose variation with x is shown in Fig. 1.

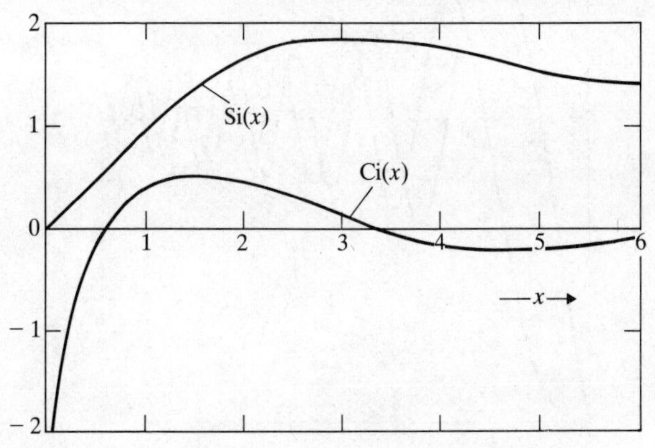

Fig. 1 Variation of Ci(x) and Si(x) with x

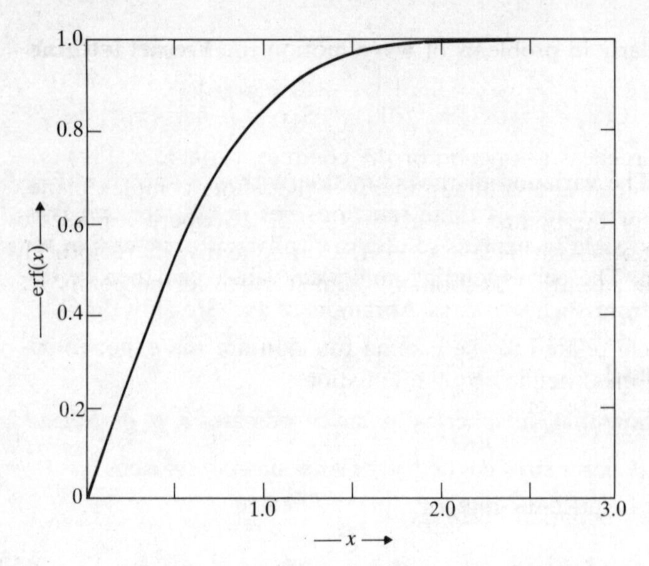

Fig. 2 Variation of erf (x) with x

In heat conduction problems solutions can often be expressed in terms of the **error-function**

$$\operatorname{erf}(x) = \frac{2}{\sqrt{\pi}} \int_0^x e^{-u^2} \, \mathrm{d}u \tag{5.11}$$

whose variation with x is exhibited graphically in Fig. 2.

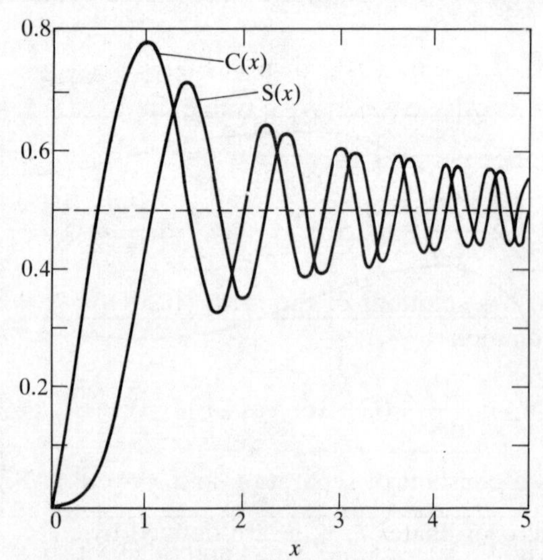

Fig. 3 Variation of the Fresnel integrals, $\mathrm{C}(x)$ and $\mathrm{S}(x)$, with x

Similarly in problems of wave motion the **Fresnel integrals**

$$C(x)=\int_0^x \cos(\tfrac{1}{2}\pi u^2)\,du, \qquad S(x)=\int_0^x \sin(\tfrac{1}{2}\pi u^2)\,du \tag{5.12}$$

occur. The variation of these function with x is shown in Fig. 3.

The importance of these functions lies in the fact that it is often possible to express solutions of physical problems in terms of them. The corresponding numerical values can then be obtained from such works as Abramowitz and Stegun (1966).

Problems I

1.1 Show that, in spherical polar coordinates r, θ, ϕ defined by

$$x=r\sin\theta\cos\phi, \quad y=r\sin\theta\sin\phi, \quad z=r\cos\theta,$$

Laplace's equation becomes

$$\frac{\partial}{\partial r}\left(r^2\frac{\partial\psi}{\partial r}\right)+\frac{1}{\sin\theta}\frac{\partial}{\partial\theta}\left(\sin\theta\frac{\partial\psi}{\partial\theta}\right)+\frac{1}{\sin^2\theta}\frac{\partial^2\psi}{\partial\phi^2}=0,$$

and prove that it possesses solutions of the form $r^n e^{im\phi}\Theta(\cos\theta)$, where $\Theta(\mu)$ satisfies the ordinary differential equation

$$(1-\mu^2)\frac{d^2\Theta}{d\mu^2}-2\mu\frac{d\Theta}{d\mu}+\left\{n(n+1)-\frac{m^2}{1-\mu^2}\right\}\Theta=0.$$

This equation is known as **Legendre's associated equation.**

1.2 Show that if

$$x=a\cosh\xi\cos\eta, \quad y=a\sinh\xi\sin\eta, \quad z=z,$$

Laplace's equation assumes the form

$$\frac{\partial^2\psi}{\partial\xi^2}+\frac{\partial^2\psi}{\partial\eta^2}+a^2(\cosh^2\xi-\cos^2\eta)\frac{\partial^2\psi}{\partial z^2}=0.$$

Deduce that it has solutions of the form $f(i\xi)f(\eta)e^{-\gamma z}$ where $f(\eta)$ satisfies the equation

$$\frac{d^2f}{d\eta^2}+(G+16q\cos 2\eta)f=0,$$

in which G is a constant of separation and $q=-a^2\gamma^2/32$.

1.3 Parabolic coordinates ξ, η, ϕ are defined by

$$x=\sqrt{(\xi\eta)}\cos\phi, \quad y=\sqrt{(\xi\eta)}\sin\phi, \quad z=\tfrac{1}{2}(\xi-\eta).$$

Show that in these coordinates Laplace's equation becomes

$$\frac{4}{\xi+\eta}\frac{\partial}{\partial \xi}\left(\xi\frac{\partial \psi}{\partial \xi}\right)+\frac{4}{\xi+\eta}\frac{\partial}{\partial \eta}\left(\eta\frac{\partial \psi}{\partial \eta}\right)+\frac{1}{\xi\eta}\frac{\partial^2 \psi}{\partial \phi^2}=0.$$

Prove that if $F_n(x)$ is a solution of the equation

$$x\frac{d^2F}{dx^2}+\frac{dF}{dx}+\left(n-\frac{m^2}{4x}\right)F=0,$$

then $F_n(\xi)F_{-n}(\eta)e^{\pm im\phi}$ is a solution of Laplace's equation.

1.4 Defining cos x, sin x to be the solutions of

$$\frac{d^2y}{dx^2}+y=0$$

which respectively are 1, 0 when $x=0$, prove

(i) $\cos(-x)=\cos x$, $\sin(-x)=-\sin x$;
(ii) $\cos(x+x')=\cos x\cos x'-\sin x\sin x'$;
(iii) $\sin(x+x')=\sin x\cos x'+\cos x\sin x'$;
(iv) $\cos^2 x+\sin^2 x=1$;
(v) $\frac{d}{dx}(\sin x)=\cos x$, $\quad\frac{d}{dx}(\cos x)=-\sin x$.

1.5 The only singularities of the differential equation

$$y''+p(x)y'+q(x)y=0$$

are regular singularities at $x=1$ of exponents α, α' and at $x=-1$ of exponents β, β' the point at infinity being an ordinary point. Prove that $\beta=-\alpha$, $\beta'=-\alpha'$ and that the differential equation is

$$(x^2-1)^2y''+2(x-1)(x-\alpha-\alpha')y'+4\alpha\alpha'y=0.$$

Show that the solution is

$$y=c_1\left(\frac{x-1}{x+1}\right)^{\alpha}+c_2\left(\frac{x-1}{x+1}\right)^{\alpha'},$$

where c_1 and c_2 are constants.

1.6 Apply the method of solution in series to the equations

$$x\frac{d^2y}{dx^2}+(a-x)\frac{dy}{dx}-y=0$$

showing that, near $x=0$, $y=Au+Bv$, where u is a Maclaurin series and $v=x^{1-a}e^x$. (a is not an integer.)

1.7 Find two solutions of the equation

$$(x^2+2x)\frac{d^2y}{dx^2}+\frac{dy}{dx}-k(k+1)y=0$$

in the form

$$y=\sum_{n=0}^{\infty} a_n x^{n+\rho}.$$

Show that, if k is a positive integer, one of these solutions is the polynomial

$$1+k(k+1)\sum_{n=1}^{k+1}\frac{(k+n-1)!\,(2x)^n}{(k-n+1)!\,(2n)!}.$$

1.8 Show that

$$w(z)=\int_C p^{-\nu-1}(p-1)^{\nu}e^{pz}\,dp$$

is a solution of the equation

$$z\frac{d^2w}{dz^2}+(1-z)\frac{dw}{dz}+\nu w=0$$

provided the contour C is such that

$$\Delta_C[p^{-\nu}(p-1)^{\nu+1}e^{pz}]=0.$$

1.9 Show that the equation

$$\frac{d^2w}{dz^2}+\tfrac{1}{3}zw(z)=0$$

has solution

$$w_1(z)=\frac{1}{2\pi i}\int_C \exp(pz+p^3)\,dp,$$

where C is an infinite contour lying wholly in the right half-plane coming from infinity in the direction $\arg p=-\frac{1}{3}\pi i$ and approaching infinity in the direction $\arg p=\frac{1}{3}\pi i$.

Distorting C continuously into the imaginary axis, show that

$$w_1(z)=2i\int_0^{\infty}\cos(tz-t^3)\,dt$$

1.10 Show that if

$$w(z)=\int_C \exp(pz^2)g(p)\,dp,$$

then

$$w''(z)=4\Delta_C[p^2g(p)\exp(pz^2)]-2\int_C \{2p^2g'(p)+3pg(p)\}\exp(pz^2)\,dp.$$

Hence show that equation (3.6) has contour integral solutions of the form

$$\int_C p^{\nu+1}\exp(pz^2-\tfrac{1}{4}p^{-1})\,dp,$$

where the contour C is such that

$$\Delta_C[p^{\nu+1}\exp(pz^2-\tfrac{1}{4}p^{-1})]=0.$$

1.11 Show that the function

$$w(z)=\int_C e^s s^{\alpha-\gamma}(s-z)^{-\alpha}\,ds$$

is a solution of the differential equation

$$z\frac{d^2w}{dz^2}+(\gamma-z)\frac{dw}{dz}-\alpha\omega=0,$$

provided that C is such that

$$\Delta_C[e^s s^{\alpha-\gamma+1}(s-z)^{-\alpha-1}]=0.$$

1.12 Show that the differential equation

$$z(1-z)\frac{d^2w}{dz^2}+\{\gamma-(\alpha+\beta+1)z\}\frac{dw}{dz}-\alpha\beta w=0$$

has solution

$$\int_C s^{\alpha-\gamma}(1-s)^{\gamma-\beta-1}(s-z)^{-\alpha}\,ds,$$

where the contour C is such that

$$\Delta_C[s^{\alpha-\gamma+1}(1-s)^{\gamma-\beta}(s-z)^{-\alpha-1}]=0.$$

1.13 Show that the differential equation

$$\frac{d^2w}{dz^2}+2z\frac{dw}{dz}+2\nu w=0$$

has solutions of the form

$$\int_C (s-z)^{-\nu} e^{-s^2}\,ds$$

provided that

$$\Delta_C[(s-z)^{-\nu} e^{-s^2}]=0.$$

Given that ν is a positive integer, $n+1$ say, deduce the solution

$$\frac{d^n}{dz^n}\, e^{-z^2}$$

1.14 Show that, if z does not lie in the segment $[-1, 1]$ of the real line, Legendre's equation (3.24) has a solution

$$-\tfrac{1}{4} i \operatorname{cosec}(\nu\pi) \int_C \frac{(s^2-1)^\nu}{(z-s)^{\nu+1}}\,ds$$

in which C is a figure-of-eight going once round the point $s=1$ in a counter-clockwise direction and once round $s=-1$ in a clockwise direction.

Distorting C into the segment $[-1+\delta, 1-\delta]$, $(\delta>0)$, taken twice and the broken circles $|s\pm 1|=\delta$, show that

$$Q_\nu(z)=\frac{1}{2^{\nu+1}}\int_{-1}^{1} \frac{(1-t^2)^\nu\,dt}{(z-t)^{\nu+1}}$$

is also a solution. Using the substitution

$$t=\frac{\sqrt{(z+1)}e^u-\sqrt{(z-1)}}{\sqrt{(z+1)}e^u+\sqrt{(z-1)}},$$

show that

$$Q_\nu(z)=\tfrac{1}{2}\int_0^\infty \{z+\sqrt{(z^2-1)}\cosh u\}^{-\nu-1}\,du.$$

1.15 Taking, in equation (3.24), $A=2^{-\nu}e^{i\nu\pi}\pi^{-1}$ and C to be $(-\infty, -1-\delta]$ taken twice together with the broken circle $|s+1|=\delta$, show that if $-1<\nu<0$ and z does not lie on the real axis between $-\infty$ and -1, Legendre's equation (3.24) has a solution

$$\frac{2^{-\nu}}{\Gamma(-\nu)\Gamma(1+\nu)}\int_0^\infty (z+\cosh t)^{-\nu-1}\sinh^{2\nu+1} t\,dt.$$

1.16 Show that if $y_1(x)$ is a solution of the self-adjoint equation (4.1) with $p>0$, then

$$y_2(x)=y_1(x)\int^x \frac{dt}{p(t)\{y_1(t)\}^2}$$

is also a solution and that

$$W(y_1, y_2; x)=\frac{1}{p(x)}.$$

1.17 Show that any solution of the equation (4.1) can be written in the form

$$y(x)=\{p(x)\}^{-\frac{1}{2}}w(x),$$

where $w(x)$ is a solution of the equation (known as the **normal form** of equation (4.1))

$$\frac{d^2w}{dx^2}+P(x)w=0,$$

in which the function P is defined by the equation

$$P(x)=\left[\frac{p'(x)}{2p(x)}\right]^2+\frac{q(x)-\frac{1}{2}p''(x)}{2p(x)}.$$

1.18 Show that the normal form of Bessel's equation (1.10) is

$$\frac{d^2y}{dx^2}+\left[1+\frac{\frac{1}{4}-\nu^2}{x^2}\right]y=0.$$

Show that if δ_ν denotes the distance between two successive zeros of any solution of (1.10), then if $|\nu|<\frac{1}{2}$, $x>\eta$, arbitrary, then

$$\pi[1+(\tfrac{1}{4}-\nu^2)/\eta^2]^{-\frac{1}{2}}<\delta_\nu<\pi,$$

and if $|\nu|>\frac{1}{2}$, $x>\eta>(\nu^2-\frac{1}{4})^{\frac{1}{2}}$, then

$$\pi<\delta_\nu<\pi[1-(\nu^2-\tfrac{1}{4})/\eta^2]^{-\frac{1}{2}}.$$

Deduce that every solution of Bessel's equation is ultimately oscillatory and possesses an infinity of zeros, the distance between any two zeros tending to π as $x\to\infty$.

1.19 A polynomial $p_n(x)$ is said to be a least square approximation to a function f defined on $[a, b]$ if it minimises the quantity

$$\|f-p_n(x)\|^2.$$

Show that if $\{\psi_r\}_{r=0}^{\infty}$ is a set of orthonormal polynomials on $[a, b]$, with p_r a polynomial of degree r, the least square approximation in this sense is the truncated Fourier series

$$\sum_{r=0}^{n} \langle f, \psi_r \rangle \psi_r(x).$$

1.20 Suppose that $\{p_n\}_{n=0}^{\infty}$ is a set of orthonormal polynomials on $[a, b]$ with respect to a weight function w. Show that if p_n is a polynomial of degree n, then there are constants A_k, B_k and C_k such that

$$p_{k+1}(x) = (A_k x + B_k)p_k(x) - C_k p_{k-1}(x), \qquad k > 1,$$

where $C_k = A_k/A_{k-1}$.

Deduce that

(i) $\displaystyle\sum_{r=0}^{n} p_r(x)p_r(y) = \frac{1}{A_n} \frac{p_{n+1}(x)p_n(y) - p_n(x)p_{n+1}(y)}{x-y}$;

(ii) $\displaystyle\sum_{r=0}^{n} p_r(x_k)p_r(x) = \frac{1}{A_n} \frac{p_{n+1}(x_k)p_n(x)}{x_k - x}$;

(iii) $\displaystyle\int_a^b \frac{p_n(x)}{x - x_k} w(x)\, dx = \frac{A_{n-1}}{p_{n-1}(x_k)}$;

where $\{x_k\}_{k=1}^{n}$ are the zeros of $p_n(x)$.

1.21 Suppose that $\{p_n\}_{n=1}^{\infty}$ is a system of polynomials orthogonal with respect to a weight function w on $[-1, 1]$. Then $p_n(x)$ has n real distinct zeros located in the interior of $[-1, 1]$.

1.22 Given that m and n are integers, evaluate

$$\frac{1}{2\pi i} \int_C \left(z - \frac{1}{z}\right)^m z^{n-1}\, dz,$$

where C is the unit circle in the z-plane.

Deduce that if r is zero or a positive integer,

$$\int_0^{2\pi} \sin^{n+2r}\theta \cos n\theta\, d\theta = \begin{cases} (-1)^{\frac{1}{2}n} \dfrac{(n+2r)!}{r!\,(n+r)!} \dfrac{\pi}{2^{n+2r-1}} & \text{if } n \text{ is even;} \\ 0, \quad \text{if } n \text{ is odd.} \end{cases}$$

$$\int_0^{2\pi} \sin^{n+2r}\theta \sin(n\theta)\, d\theta = \begin{cases} 0, \quad \text{if } n \text{ is even;} \\ (-1)^{\frac{1}{2}n-\frac{1}{2}} \dfrac{(n+2r)!}{r!\,(n+r)!} \dfrac{\pi}{2^{n+2r-1}}, & \text{if } n \text{ is odd.} \end{cases}$$

State the values of these integrals when r is a negative integer such that $r > -\frac{1}{2}(n+1)$.

Derive the corresponding results for the integrals

$$\int_0^{2\pi} \sin^{n+2r+1}\theta \cos(n\theta)\, d\theta, \qquad \int_0^{2\pi} \sin^{n+2r+1}\theta \sin(n\theta)\, d\theta.$$

1.23 Show that if m and n are positive integers (or zero) and $m > n$, then

$$\frac{1}{2\pi i}\int_C \left(z+\frac{1}{z}\right)^{2m} z^{2n-1}\, dz = \frac{(2m)!}{(m+n)!\,(m-n)!},$$

where C is the unit circle $|z| = 1$.

Deduce that

$$\int_0^{\frac{1}{2}\pi} \cos^{2m}\theta \cos(2n\theta)\, d\theta = \frac{(2m)!}{(m+n)!\,(m-n)!}\frac{\pi}{2^{2m+1}}, \qquad m > n.$$

1.24 Prove that if s is an integer and a is fractional, then

$$\Gamma(a-s) = (-1)^s \frac{\Gamma(a)\Gamma(1-a)}{\Gamma(1-a+s)}.$$

1.25 Prove that if z is not an integer

$$\Gamma(z) = (e^{2\pi i z}-1)^{-1}\int_C e^{-s}s^{z-1}\, ds, \qquad \text{Re } z > 0$$

where the path of integration C starts at $+\infty$ on the real axis, encircles the origin in the counter-clockwise direction and returns to the starting point.

Deduce that

(i) $\Gamma(z) = \dfrac{i}{2\sin(\pi z)}\displaystyle\int_C e^{-s}(-s)^{z-1}\, ds;$

(ii) $\dfrac{1}{\Gamma(z)} = \dfrac{1}{2\pi}\displaystyle\int_C e^{-s}(-s)^{-z}\, ds.$

1.26 Prove that if $a > 0$, $b > 0$,

$$B(a, b) = (1-e^{2\pi i a})^{-1}(1-e^{2\pi i b})^{-1}\int^{(1+,\,0+,\,1-,\,0-)} s^{a-1}(1-s)^{b-1}\, ds,$$

where the notation indicates that the integral is taken over a path in the s-plane that encircles the points $s = 1$, $s = 0$ first in the counter-clockwise direction and then in the clockwise direction.

1.27 The **Pochhammer symbol** $(\alpha)_n$ is defined for a positive integer n by the equation

$$(\alpha)_n = \alpha(\alpha+1)\ldots(\alpha+n-1) = \Gamma(\alpha+n)/\Gamma(\alpha).$$

Prove that if m, n, r and s are positive integers

(i) $(\alpha-1)(\alpha)_{n-1} = (\alpha-1)_n$;

(ii) $(\alpha)_{n-r} = (-1)^r \dfrac{(\alpha)_n}{(1-\alpha-n)_r}$;

(iii) $\dfrac{n!}{(n-s)!} = (-1)^s(-n)_s, \qquad (n>s)$;

(iv) $(-m)_r(-m+r)_s = (-m)_s(-m+s)_r$;

(v) $(-m)_{m-s} = (-1)^{m-s}\dfrac{m!}{s!}$;

(vi) $(\frac{1}{2}-2m)_{m-s} = \dfrac{\Gamma(2m+\frac{1}{2})}{\Gamma(m+\frac{1}{2})}\dfrac{(-1)^{m-s}}{(m+\frac{1}{2})_s}$.

1.28 Prove that

(i) $\operatorname{erf}(z) = \dfrac{2}{\sqrt{\pi}}\displaystyle\sum_{r=0}^{\infty}\dfrac{(-1)^r}{r!}\dfrac{z^{2r+1}}{2r+1}$;

(ii) $\displaystyle\int_0^\infty \frac{\sin(xt)}{t} e^{-t^2}\,dt = \tfrac{1}{2}\pi \operatorname{erf}(\tfrac{1}{2}x), \qquad x>0.$

1.29 Prove that, if $x>0$

(i) $\displaystyle\int_0^1 \frac{e^{-x^2(1+t^2)}}{1+t^2}\,dt = \tfrac{1}{4}\pi[1-\{\operatorname{erf}(x)\}^2]$;

(ii) $\displaystyle\int_1^\infty \frac{e^{-x^2(1+t^2)}}{1+t^2}\,dt = \tfrac{1}{4}\pi\{\operatorname{erfc}(x)\}^2$.

1.30 Prove that

$$\operatorname{ei}(x) = -\gamma - \log x + x - \frac{x^2}{2.2!} + \frac{x^3}{3.3!} - \ldots$$

and deduce that

$$\operatorname{Ci}(x) = \gamma + \log x - \frac{x^2}{2.2!} + \frac{x^4}{4.4!} - \ldots$$

$$\operatorname{Si}(x) = x - \frac{x^3}{3.3!} + \frac{x^5}{5.5!} - \ldots.$$

2
Hypergeometric functions

§6. The hypergeometric series

The series

$$1+\frac{\alpha\beta}{1.\gamma}x+\frac{\alpha(\alpha+1)\beta(\beta+1)}{1.2\gamma(\gamma+1)}x^2+\ldots \tag{6.1}$$

is of great importance in mathematics. Since it is an obvious generalisation of the geometric series

$$1+x+x^2+\ldots,$$

it is called the **hypergeometric series.** It is readily shown that, provided γ is not zero or a negative integer, the series is absolutely convergent if $|x|<1$, divergent if $|x|>1$, while if $|x|=1$ the series converges absolutely if $\gamma>\alpha+\beta$. It is convergent when $x=-1$, provided that $\gamma>\alpha+\beta-1$.

If we introduce the notation

$$(\alpha)_r=\alpha(\alpha+1)\ldots(\alpha+r-1)=\frac{\Gamma(\alpha+r)}{\Gamma(\alpha)} \tag{6.2}$$

we may write the series (6.1) in the form

$${}_2F_1(\alpha,\beta;\gamma;x)=\sum_{r=0}^{\infty}\frac{(\alpha)_r(\beta)_r}{r!\,(\gamma)_r}x^r, \tag{6.3}$$

the suffixes 2 and 1 denoting that there are two parameters of the type α and one of the type γ. We shall generalise this concept at a later stage (§12 below) but it is advisable at this stage to denote the 'ordinary' hypergeometric function by the symbol ${}_2F_1$ instead of simply F, if we are to avoid confusion later. From the definition (6.3) it is obvious that

$${}_2F_1(\beta,\alpha;\gamma;x)={}_2F_1(\alpha,\beta;\gamma;x). \tag{6.4}$$

A significant property of the hypergeometric series follows immediately from the definition (6.3). We have

$$\frac{d}{dx}\,{}_2F_1(\alpha, \beta; \gamma; x) = \sum_{r=1}^{\infty} \frac{(\alpha)_r(\beta)_r}{(r-1)!(\gamma)_r} x^{r-1}$$
$$= \sum_{r=0}^{\infty} \frac{(\alpha)_{r+1}(\beta)_{r+1}}{r!(\gamma)_{r+1}} x^r.$$

Now $(\alpha)_{r+1} = \alpha(\alpha+1)_r$ so the right-hand side of the last equation becomes

$$\frac{\alpha\beta}{\gamma} \sum_{r=0}^{\infty} \frac{(\alpha+1)_r(\beta+1)_r}{r!(\gamma+1)_r} x^r$$

showing that

$$\frac{d}{dx}\,{}_2F_1(\alpha, \beta; \gamma; x) = \frac{\alpha\beta}{\gamma}\,{}_2F_1(\alpha+1, \beta+1; \gamma+1; x). \tag{6.5}$$

It should also be observed that

$${}_2F_1(\alpha, \beta; \gamma; 0) = 1 \tag{6.6}$$

so that

$$\left[\frac{d}{dx}\,{}_2F_1(\alpha, \beta; \gamma; x)\right]_{x=0} = \frac{\alpha\beta}{\gamma}. \tag{6.7}$$

Several well-known elementary functions can be expressed as hypergeometric series; examples of them are given in Problem 2.1.

It should be noted that, if we adopt a certain convention, a hypergeometric series can stop and start again after a number of zero terms. For example, consider the hypergeometric series ${}_2F_1(-n; b; -n-m; x)$ where both m and n are positive integers and b is neither zero nor a negative integer. Because of the occurrence of $(-n)_r$ in the numerator in the expansion in powers of x it is obvious that the $(n+1)$th term of the expansion will be zero, and we are tempted to think that every subsequent term is also zero. If we note that, as a result of Problem 1.27(iii)

$$\frac{(-n)_r}{(-n-m)_r} = \frac{n!}{(n+m)!}(n+m-r)(n+m-r-1)\ldots(n-r+1) \tag{6.8}$$

when the form on the left is not of type 0/0, and if, further, we assume that it still has the value on the left when it is indeterminate, we see that we may write

$$
\begin{aligned}
&{}_2F_1(-n, b; -n-m; x) \\
&\quad = \sum_{r=0}^{\infty} \left(1-\frac{r}{n+m}\right)\left(1-\frac{r}{n+m-1}\right)\ldots\left(1-\frac{r}{n+1}\right)\frac{(b)_r x^r}{r!}, \qquad (6.9)
\end{aligned}
$$

so that although the series stops at the nth term it starts up again at the $(n+m+1)$th term. For instance,

$$
{}_2F_1(-2, 1; -5; x) = 1 + \tfrac{2}{5}x + \tfrac{1}{10}x^2 - \tfrac{1}{10}x^6 - \tfrac{2}{5}x^7 - x^8 + \ldots
$$

According to an alternative convention, however, the hypergeometric function does not restart after a set of zero terms.

§7. An integral formula for the hypergeometric series

In order to derive some further properties of the hypergeometric series we shall first of all establish an expression for the series in the form of an integral. It is readily shown that

$$
\frac{(\beta)_r}{(\gamma)_r} = \frac{B(\beta+r, \gamma-\beta)}{B(\beta, \gamma-\beta)} = \frac{1}{B(\beta, \gamma-\beta)} \int_0^1 (1-t)^{\gamma-\beta-1} t^{\beta+r-1}\, dt
$$

from which it follows that

$$
{}_2F_1(\alpha, \beta; \gamma; x) = \frac{1}{B(\beta, \gamma-\beta)} \sum_{r=0}^{\infty} \frac{(\alpha)_r}{r!} x^r \int_0^1 (1-t)^{\gamma-\beta-1} t^{\beta+r-1}\, dt.
$$

Interchanging the order in which the operations of summation and integration are performed we see that

$$
{}_2F_1(\alpha, \beta; \gamma; x) = \frac{1}{B(\beta, \gamma-\beta)} \int_0^1 (1-t)^{\gamma-\beta-1} t^{\beta-1} \left\{ \sum_{r=0}^{\infty} \frac{(\alpha)_r}{r!} (xt)^r \right\} dt.
$$

Using the fact that

$$
\sum_{r=0}^{\infty} \frac{(\alpha)_r}{r!} (xt)^r = (1-xt)^{-\alpha},
$$

we have the integral formula

$$
{}_2F_1(\alpha, \beta; \gamma; x) = \frac{1}{B(\beta, \gamma-\beta)} \int_0^1 (1-t)^{\gamma-\beta-1} t^{\beta-1} (1-xt)^{-\alpha}\, dt, \qquad (7.1)
$$

valid if $|x|<1$, $\gamma>\beta>0$. The results hold if x is complex provided that we choose the branch of $(1-xt)^{-\alpha}$ in such a way that $(1-xt)^{-\alpha}\to 1$ as $t\to 0$ and $\mathcal{R}(\gamma)>\mathcal{R}(\beta)>0$.

The first application of (7.1) is the derivation of the value of the hypergeometric series with unit argument. Putting $x=1$ in (7.1) we have

$$
\begin{aligned}
{}_2F_1(\alpha, \beta; \gamma; 1) &= \frac{1}{B(\beta, \gamma-\beta)}\int_0^1 (1-t)^{\gamma-\alpha-\beta-1}t^{\beta-1}\,\mathrm{d}t \\
&= \frac{B(\beta, \gamma-\alpha-\beta)}{B(\beta, \gamma-\beta)}
\end{aligned}
$$

if $\gamma-\alpha-\beta>0$, $\beta>0$. If we express the beta function in terms of gamma functions we have **Gauss's theorem:**

$$
{}_2F_1(\alpha, \beta; \gamma; 1) = \frac{\Gamma(\gamma)\Gamma(\gamma-\alpha-\beta)}{\Gamma(\gamma-\alpha)\Gamma(\gamma-\beta)}. \tag{7.2}
$$

Now if $\alpha=-n$, a negative integer, we have

$$
\frac{\Gamma(\gamma-\alpha-\beta)}{\Gamma(\gamma-\beta)} = (\gamma-\beta)_n, \qquad \frac{\Gamma(\gamma-\alpha)}{\Gamma(\gamma)} = (\gamma)_n
$$

so that equation (7.2) reduces to

$$
{}_2F_1(-n, \beta; \gamma; 1) = \frac{(\gamma-\beta)_n}{(\gamma)_n}
$$

which is known, in elementary mathematics, as **Vandermonde's theorem.**

Again, if we put $x=-1$ and $\alpha=1+\beta-\gamma$ we have, from equation (7.1)

$$
{}_2F_1(\alpha, \beta; \beta-\alpha+1; -1) = \frac{\Gamma(1+\beta-\alpha)}{\Gamma(\beta)\Gamma(1-\alpha)}\int_0^1 (1-t^2)^{-\alpha}t^{\beta-1}\,\mathrm{d}t.
$$

If we write $\xi=t^2$ in this integral we see that its value is $\frac{1}{2}B(\frac{1}{2}\beta, 1-\alpha)$. Using this result and the relation $\frac{1}{2}\Gamma(\frac{1}{2}\beta)/\Gamma(\beta)=\Gamma(1+\frac{1}{2}\beta)/\Gamma(1+\beta)$ we have **Kummer's theorem**

$$
{}_2F_1(\alpha, \beta, \beta-\alpha+1; -1) = \frac{\Gamma(1+\beta-\alpha)\Gamma(1+\frac{1}{2}\beta)}{\Gamma(1+\beta)\Gamma(1+\frac{1}{2}\beta-\alpha)}. \tag{7.3}
$$

Further, we can deduce from the formula (7.1) relations between hypergeometric series of argument x and those of argument

$x/(x-1)$. Putting $\tau = 1 - t$ in equation (7.1), and noting that

$$\{1 - x(1-\tau)\}^{-\alpha} = (1-x)^{-\alpha}\left\{1 - \frac{x}{x-1}\tau\right\}^{-\alpha},$$

we see that

$$\begin{aligned} {}_2F_1(\alpha, \beta; \gamma; x) &= \frac{(1-x)^{-\alpha}}{B(\beta, \gamma-\beta)} \int_0^1 (1-\tau)^{\beta-1}\tau^{\gamma-\beta-1}\left\{1 - \frac{x}{x-1}\tau\right\}^{-\alpha} d\tau \\ &= \frac{(1-x)^{-\alpha}}{B(\beta, \gamma-\beta)} B(\gamma-\beta, \beta)\, {}_2F_1\left(\alpha, \gamma-\beta; \gamma; \frac{x}{x-1}\right), \end{aligned}$$

whence we have the relation

$$ {}_2F_1(\alpha, \beta; \gamma; x) = (1-x)^{-\alpha}\, {}_2F_1\left(\alpha, \gamma-\beta; \gamma; \frac{x}{x-1}\right), \tag{7.4}$$

and, by symmetry, the relation

$$ {}_2F_1(\alpha, \beta; y; x) = (1-x)^{-\beta}\, {}_2F_1\left(\gamma-\alpha, \beta; \gamma; \frac{x}{x-1}\right). \tag{7.5}$$

Using the symmetry relation (6.4) and equation (7.4) with x replaced by $x/(x-1)$ we see that

$$\begin{aligned} {}_2F_1\left(\alpha, \gamma-\beta; \gamma; \frac{x}{x-1}\right) &= {}_2F_1\left(\gamma-\beta, \alpha; \gamma; \frac{x}{x-1}\right) \\ &= (1-x)^{\gamma-\beta}\, {}_2F_1(\gamma-\beta, \gamma-\alpha; \gamma; x), \end{aligned}$$

so that

$$ {}_2F_1(\alpha, \beta; \gamma; x) = (1-x)^{\gamma-\alpha-\beta}\, {}_2F_1(\gamma-\alpha, \gamma-\beta; \gamma; x). \tag{7.6}$$

If we put $x = \frac{1}{2}$ in equation (7.4) we obtain the relation

$$ {}_2F_1(\alpha, \beta; \gamma; \tfrac{1}{2}) = 2^{\alpha}\, {}_2F_1(\alpha, \gamma-\beta; \gamma; -1).$$

The series on the right-hand side of this equation can be derived from equation (7.3) provided either that

$$\gamma = \gamma - \beta - \alpha + 1, \quad \text{i.e.} \quad \beta = 1 - \alpha,$$

or that

$$\gamma = \alpha - (\gamma - \beta) + 1, \quad \text{i.e.} \quad \gamma = \tfrac{1}{2}(\alpha + \beta + 1).$$

We then obtain the formulae

$$_2F_1(\alpha, 1-\alpha; \gamma; \tfrac{1}{2}) = \frac{\Gamma(\frac{1}{2}\gamma)\Gamma(\frac{1}{2}\gamma+\frac{1}{2})}{\Gamma(\frac{1}{2}\alpha+\frac{1}{2}\gamma)\Gamma(\frac{1}{2}-\frac{1}{2}\alpha+\frac{1}{2}\gamma)}; \tag{7.7}$$

$$_2F_1(\alpha, \beta; \tfrac{1}{2}\alpha+\tfrac{1}{2}\beta+\tfrac{1}{2}; \tfrac{1}{2}) = \frac{\Gamma(\frac{1}{2})\Gamma(\frac{1}{2}+\frac{1}{2}\alpha+\frac{1}{2}\beta)}{\Gamma(\frac{1}{2}+\frac{1}{2}\alpha)\Gamma(\frac{1}{2}+\frac{1}{2}\beta)}. \tag{7.8}$$

§8. The hypergeometric equation

In certain problems it is possible to reduce the solution to that of solving the second order linear differential equation

$$x(1-x)\frac{d^2y}{dx^2}+\{\gamma-(1+\alpha+\beta)x\}\frac{dy}{dx}-\alpha\beta y=0 \tag{8.1}$$

in which α, β and γ are constants. For instance, the Schrödinger equation for a symmetrical-top molecule, which is of importance in the theory of molecular spectra, can, by simple transformations, be reduced to this type. An equation of this type also arises in the study of the flow of compressible fluids. In addition certain other differential equations (such as that occurring in Problem 1.1.) which arise in the solution of boundary value problems in mathematical physics can, by a simple change of variable, be transformed to an equation of type (8.1). Indeed it can be shown that any ordinary linear differential equation of the second order whose only singular points are regular singular points, one of which may be the point at infinity, can be transformed to the form (8.1). For that reason it is desirable to investigate the nature of the solutions of this equation, which is called the **hypergeometric equation.**

We may write the hypergeometric equation in the form

$$x^2y''+x(1+x+x^2+\ldots)\{\gamma-(\alpha+\beta+1)x\}y' \\ -\alpha\beta x(1+x+x^2+\ldots)y=0,$$

so that, in the notation of §2.2, we see that near $x=0$

$$p_0=\gamma, \qquad q_0=0,$$

and the indicial equation is

$$\rho^2+(\gamma-1)\rho=0$$

with roots $\rho=0$ and $\rho=1-\gamma$.

Similarly, the equation can be put in the form

$$(x-1)^2y''-(x-1)\{\gamma-\alpha-\beta-1-\gamma(x-1)+\ldots\}y'$$
$$+\alpha\beta(x-1)\{1-(x-1)+\ldots\}y=0,$$

with indicial equation

$$\rho^2+(\alpha+\beta-\gamma)\rho=0,$$

of which the roots are $\rho=0$, $\rho=\gamma-\alpha-\beta$.

Finally in the notation of §4 we have for large values of x

$$\alpha(x)\sim\frac{(\alpha+\beta+1)}{x}, \qquad \beta(x)\sim\frac{\alpha\beta}{x^2}$$

and so the indicial equation appropriate to the point at infinity is

$$\rho^2-(\alpha+\beta)\rho+\alpha\beta=0,$$

with roots α, β.

Thus the regular singular points of the hypergeometric equation are:

(i) $x=0$ with exponents 0, $1-\gamma$;
(ii) $x=\infty$ with exponents α, β;
(iii) $x=1$ with exponents 0, $\gamma-\alpha-\beta$.

These facts are exhibited symbolically by denoting the most general solution of the hypergeometric equation by a scheme of the form

$$y=P\left\{\begin{matrix}0 & \infty & 1 & \\ 0 & \alpha & 0 & x \\ 1-\gamma & \beta & \gamma-\alpha-\beta & \end{matrix}\right\}. \tag{8.2}$$

The symbol on the right is called the **Riemann-*P*-function** of the equation.

We shall now consider the form of the solutions in the neighbourhood of the regular singular points.

(a) $x=0$: Corresponding to the root $\rho=0$ we have a solution of the form

$$y=\sum_{r=0}^{\infty}c_rx^r.$$

Substituting this series into equation (8.1) we obtain the relation

$$(1-x)\sum_{r=0}^{\infty} c_r r(r-1)x^{r-1}+\{\gamma-(\alpha+\beta+1)x\}\sum_{r=0}^{\infty} c_r r x^{r-1}-\alpha\beta\sum_{r=0}^{\infty} c_r x^r=0,$$

which is readily seen to be equivalent to

$$\sum_{r=0}^{\infty}\{c_{r+1}[r(r+1)+(r+1)\gamma]-c_r(r+\alpha)(r+\beta)\}x^r=0,$$

so that

$$c_{r+1}=\frac{(r+\alpha)(r+\beta)}{(r+1)(r+\gamma)}c_r, \tag{8.3}$$

from which it follows that

$$c_r=\frac{(\alpha)_r(\beta)_r}{(\gamma)_r r!}c_0. \tag{8.4}$$

It follows that the solution which reduces to unity when $x=0$ is

$$y=1+\frac{\alpha\beta}{\gamma . 1!}x+\frac{\alpha(\alpha+1)\beta(\beta+1)}{\gamma(\gamma+1)2!}x^2+\dots$$

i.e.

$$y={}_2F_1(\alpha,\beta;\gamma;x). \tag{8.5}$$

Similarly, if $1-\gamma$ is neither zero nor a negative integer, the solution corresponding to the root $\rho=1-\gamma$ is

$$y=\sum_{r=0}^{\infty} c_r x^{1-\gamma+r},$$

where

$$(1-x)\sum_{r=0}^{\infty} c_r(r+1-\gamma)(r-\gamma)x^{r-\gamma}$$
$$+\{\gamma-(\alpha+\beta+1)x\}\sum_{r=0}^{\infty} c_r(r+1-\gamma)x^{r-\gamma}-\alpha\beta\sum_{r=0}^{\infty} c_r x^{1-\gamma+r}=0,$$

which is equivalent to

$$\sum_{r=0}^{\infty} c_r\{(r+1-\gamma)(r-\gamma)+\gamma(r+1-\gamma)\}x^{r-\gamma}$$
$$-\sum_{r=0}^{\infty} c_r\{(r+1-\gamma)(r-\gamma)+(\alpha+\beta+1)(r+1-\gamma)+\alpha\beta\}x^{r-\gamma+1}=0,$$

implying that

$$c_{r+1}=\frac{(r+\alpha-\gamma+1)(r+\beta-\gamma+1)}{(r+1)(r+2-\gamma)}c_r.$$

Comparing this relation with (8.3) and taking $c_0=1$ we see that this solution is

$$x^{1-\gamma}\,{}_2F_1(\alpha-\gamma+1,\beta-\gamma+1;2-\gamma;x). \tag{8.6}$$

Combining equations (8.5) and (8.6) we see that the general solution valid in the neighbourhood of the origin is

$$y=A\,{}_2F_1(\alpha,\beta;\gamma;x) \\ +Bx^{1-\gamma}\,{}_2F_1(\alpha-\gamma+1,\beta-\gamma+1;2-\gamma;x), \tag{8.7}$$

provided that $1-\gamma$ is not zero or a positive integer.

If $\gamma=1$, the solutions (8.5) and (8.6) are identical. If we write

$$y_1(x)={}_2F_1(\alpha,\beta;\gamma;x)$$

and put

$$y_2(x)=y_1(x)\log x+\sum_{r=1}^{\infty}c_rx^r,$$

we find on substituting in (8.1), with $\gamma=1$, that

$$(r+1)^2c_{r+1}-r(\alpha+\beta+1)c_r+\frac{(\alpha)_r(\beta)_r(\alpha\beta-\alpha-\beta-r)}{r!\,(r+1)!}=0$$

from which the coefficients c_r may be determined.

A similar procedure holds when $1-\gamma$ is a positive integer.

(b) $x=1$: If we let $\xi=1-x$, equation (8.1) reduces to

$$\xi(1-\xi)\frac{d^2y}{d\xi^2}+\{\alpha+\beta-\gamma+1-(\alpha+\beta+1)\xi\}\frac{dy}{d\xi}-\alpha\beta y=0$$

which is identical with equation (8.1) with γ replaced by $\alpha+\beta-\gamma+1$, and x by $\xi=1-x$. Hence it follows from equation (8.7) that the required solution is

$$y=A\,{}_2F_1(\alpha,\beta;\alpha+\beta-\gamma+1;1-x) \\ +B(1-x)^{\gamma-\alpha-\beta}\,{}_2F_1(\gamma-\alpha,\gamma-\beta;\gamma-\alpha-\beta+1;1-x). \tag{8.8}$$

(c) $x=\infty$: Corresponding to the root $\rho=\alpha$, we put

$$y=\sum_{r=0}^{\infty}c_rx^{-\alpha-r},$$

which gives

$$(1-x)\sum_{r=0}^{\infty} c_r(r+\alpha)(r+\alpha+1)x^{-r-\alpha-1}$$
$$-\{\gamma-(\alpha+\beta+1)x\}\sum_{r=0}^{\infty}(r+\alpha)c_r x^{-r-\alpha-1}-\alpha\beta\sum_{r=0}^{\infty}c_r x^{-r-\alpha}=0,$$

i.e.,

$$\sum_{r=0}^{\infty} c_r(r+\alpha)(r+\alpha-\gamma+1)x^{-r-\alpha-1}=\sum_{r=0}^{\infty} c_r r(r+\alpha-\beta)x^{-r-\alpha},$$

whence it follows that

$$c_{r+1}=\frac{(r+\alpha)(r+\alpha-\gamma+1)}{(r+1)(r+\alpha-\beta+1)}c_r,$$

which in turn is equivalent to

$$c_r=\frac{(\alpha)_r(\alpha-\gamma+1)_r}{r!\,(\alpha-\beta+1)_r}c_0.$$

Taking $c_0=1$ we obtain the solution

$$x^{-\alpha}\,{}_2F_1\left(\alpha,\alpha-\gamma+1;\alpha-\beta+1;\frac{1}{x}\right).$$

From the symmetry we see that the other solution is

$$x^{-\beta}\,{}_2F_1\left(\beta,\beta-\gamma+1;\beta-\alpha+1;\frac{1}{x}\right),$$

so that the required solution is

$$\begin{aligned}y=Ax^{-\alpha}\,{}_2F_1\left(\alpha,\alpha-\gamma+1;\alpha-\beta+1;\frac{1}{x}\right)\\+Bx^{-\beta}\,{}_2F_1\left(\beta,\beta-\gamma+1;\beta-\alpha+1;\frac{1}{x}\right).\end{aligned}\tag{8.9}$$

We can also derive contour integral expressions for the solution of the hypergeometric equation. We can write the equation (8.1) in the form (3.8) by taking

$$p(z)=z(1-z),\qquad q(s)=\gamma-(1+\alpha+\beta)z,\qquad c=\alpha\beta.$$

In this case the quadratic equation for λ is

$$\lambda^2+(\alpha+\beta)\lambda+\alpha\beta=0,$$

and the equation for the linear function L is

$$L(z)=(2\lambda+\alpha+\beta-1)z-(\lambda-1+\gamma).$$

Taking $\lambda=-\alpha$, we find that

$$L(z)=(\alpha-\gamma+1)z-(\alpha-\beta+1),$$

so that equation (3.13) for the determination of f takes the form

$$\frac{f'(s)}{f(s)}=\frac{\alpha-\gamma}{s}-\frac{\gamma-\beta-1}{1-s};$$

we may therefore take

$$f(s)=As^{\alpha-\gamma}(1-s)^{\gamma-\beta-1},$$

where A is an arbitrary constant.

In this way we obtain the solution

$$w(z)=A\int_C s^{\alpha-\gamma}(1-s)^{\gamma-\beta-1}(s-z)^{-\alpha}\,ds \tag{8.10}$$

where the contour C is such that

$$\Delta_C[s^{\alpha-\gamma+1}(1-s)^{\gamma-\beta}(s-z)^{-\alpha-1}]=0. \tag{8.11}$$

On the other hand if we take $\lambda=-\beta$ we obtain the solution

$$w(z)=A\int_C s^{\beta-\gamma}(1-s)^{\gamma-\alpha-1}(s-z)^{-\beta}\,ds, \tag{8.12}$$

where C is a contour satisfying

$$\Delta_C[s^{\beta-\gamma+1}(1-s)^{\gamma-\alpha}(s-z)^{-\beta-1}]=0. \tag{8.13}$$

The choice of C can be made from one of the possibilities (a), (b) and (c) listed in §3.2.

A third integral can be obtained by replacing s by $1/s$ in either (8.10) or (8.11); for instance, from (8.10) we derive the integral solution

$$w(z)=A\int_C s^{\beta-1}(1-s)^{\gamma-\beta-1}(1-sz)^{-\alpha}\,ds, \tag{8.14}$$

where the contour C satisfies the condition

$$\Delta_C[s^{\beta}(1-s)^{\gamma-\beta}(1-sz)^{-\alpha-1}]=0. \tag{8.15}$$

For example, we could take for C in (8.14) the contour shown in Fig. 4 which encircles the point $s=1$ in the counter-clockwise

Fig. 4

direction, then $s=0$ in the counterclockwise direction and then $s=1$ in the clockwise direction; finally it encircles $s=0$ in the clockwise direction. The convention for denoting such a contour is

$$\int^{(1+,\,0+,\,1-,\,0-)},$$

so we have the contour integral solution

$$w_1(z)=\int^{(1+,\,0+,\,1-,\,0-)} s^{\beta-1}(1-s)^{\gamma-\beta-1}(1-zs)^{-\alpha}\,\mathrm{d}s \tag{8.16}$$

of the hypergeometric equation. Deforming the path of integration into four traverses of the segment $[-1+\delta, 1-\delta]$ of the real axis and two circuits each of the circles $|z+1|=\delta$ and $|z-1|=\delta$ taken in the appropriate directions we see that

$$w_1(z)=(1-e^{2\pi i\beta})\{1-e^{2\pi i(\gamma-\beta)}\}\int_0^1 t^{\beta-1}(1-t)^{\gamma-\beta-1}(1-zt)^{-\alpha}\,\mathrm{d}t, \tag{8.17}$$

where we have assumed that $|zt|<1$ and chosen the branch of $(1-zt)^{-\alpha}$ which takes the value $+1$ when $z=0$. That the solution (8.17) is a multiple of ${}_2F_1(\alpha,\beta;\gamma;z)$ follows immediately from equation (7.1).

If we consider the integral

$$w_2(z)=\int^{(0+,\,z+,\,0-,\,z-)} s^{\alpha-\gamma}(1-s)^{\gamma-\beta-1}(z-s)^{-\alpha}\,\mathrm{d}s \tag{8.18}$$

obtained by taking a special form for C in (8.10), where the initial point is on the straight line joining $s=0$ to $s=z$ and $\arg(s/z)$ and $\arg(1-s/z)$ are both taken to be zero at that point, we find that the integral can be reduced to

$$-\{1-e^{2\pi i(\alpha-\gamma)}\}(1-e^{-2\pi i\alpha})z^{1-\gamma}\int_0^1 t^{\alpha-\gamma}(1-t)^{-\alpha}(1-zt)^{-\beta+\gamma-1}\,\mathrm{d}t.$$

Again using (7.1) we see that if $|z|<1$

$$w_2(z)=-\{1-e^{2\pi i(\gamma-\alpha)}\}(1-e^{-2\pi i\alpha})$$
$$\times B(\alpha-\gamma+1, 1-\alpha)_2F_1(\beta-\gamma+1, \alpha-\gamma+1; 2-\gamma; z)z^{1-\gamma},$$

so that $w_2(z)$ gives an expression for the function

$$z^{1-\gamma}\,{}_2F_1(\alpha-\gamma+1, \beta-\gamma+1; 2-\gamma; z)$$

which is valid for all values of z.

§9. Linear relations between the solutions of the hypergeometric equation

The series in the solution (8.7) are convergent if $|x|<1$, i.e. in the interval $(-1, 1)$ whereas those in the solution (8.8) are convergent in $(0, 2)$. There is therefore an interval, namely $(0, 1)$, in which all four series converge, and since only two solutions of the differential equation are linearly independent it follows that there must be a linear relation, valid if $0<x<1$, between solutions of type (8.7) and those of type (8.8).

Let

$${}_2F_1(\alpha, \beta; \gamma; x)=A\,{}_2F_1(\alpha, \beta; \alpha+\beta-\gamma+1; 1-x)$$
$$+B(1-x)^{\gamma-\alpha-\beta}\,{}_2F_1(\gamma-\alpha, \gamma-\beta; \gamma-\alpha-\beta+1; 1-x),$$

then putting $x=0$ we have

$$1=A\,{}_2F_1(\alpha, \beta; \alpha+\beta-\gamma+1; 1)$$
$$+B\,{}_2F_1(\gamma-\alpha, \gamma-\beta; \gamma-\alpha-\beta+1; 1),$$

and putting $x=1$ we have

$${}_2F_1(\alpha, \beta; \gamma; 1)=A,$$

if we assume that

$$1>\gamma>\alpha+\beta. \tag{9.1}$$

Substituting for the series with unit argument from equation (7.2) we see that

$$A=\frac{\Gamma(\gamma)\Gamma(\gamma-\alpha-\beta)}{\Gamma(\gamma-\alpha)\Gamma(\gamma-\beta)},$$

and that

$$1 = A\frac{\Gamma(\alpha+\beta-\gamma+1)\Gamma(1-\gamma)}{\Gamma(\beta-\gamma+1)\Gamma(\alpha-\gamma+1)} + B\frac{\Gamma(\gamma-\alpha-\beta+1)\Gamma(1-\gamma)}{\Gamma(1-\beta)\Gamma(1-\alpha)},$$

so that

$$B = \frac{\Gamma(\gamma)\Gamma(\alpha+\beta-\gamma)}{\Gamma(\alpha)\Gamma(\beta)},$$

whence we find that

$$\begin{aligned}{}_2F_1(\alpha, \beta; \gamma; x) = {} & \frac{\Gamma(\gamma)\Gamma(\gamma-\alpha-\beta)}{\Gamma(\gamma-\alpha)\Gamma(\gamma-\beta)}\,{}_2F_1(\alpha, \beta; \alpha+\beta-\gamma+1; 1-x) \\ & + \frac{\Gamma(\gamma)\Gamma(\alpha+\beta-\gamma)}{\Gamma(\alpha)\Gamma(\beta)}(1-x)^{\gamma-\alpha-\beta} \\ & \times {}_2F_1(\gamma-\alpha, \gamma-\beta; \gamma-\alpha-\beta+1; 1-x), \qquad (9.2)\end{aligned}$$

provided that the condition (9.1) is satisfied and $0 < x < 1$.

If we replace x by $1/x$ in equation (9.2) we have

$$\begin{aligned}{}_2F_1\left(\alpha, \beta; \gamma; \frac{1}{2x}\right) = {} & \frac{\Gamma(\gamma)\Gamma(\gamma-\alpha-\beta)}{\Gamma(\gamma-\alpha)\Gamma(\gamma-\beta)}\,{}_2F_1\left(\alpha, \beta; \alpha+\beta-\gamma+1; 1-\frac{1}{x}\right) \\ & + \frac{\Gamma(\gamma)\Gamma(\alpha+\beta-\gamma)}{\Gamma(\alpha)\Gamma(\beta)}\left(1-\frac{1}{x}\right)^{\gamma-\alpha-\beta} \\ & \times {}_2F_1\left(\gamma-\alpha, \gamma-\beta; \gamma-\alpha-\beta+1; 1-\frac{1}{x}\right).\end{aligned}$$

and from equation (7.4)

$${}_2F_1\left(\alpha, \beta; \gamma; 1-\frac{1}{x}\right) = x^{\alpha}\,{}_2F_1(\alpha, \gamma-\beta; \gamma; 1-x),$$

so that

$$\begin{aligned}F\left(\alpha, \beta; \gamma; \frac{1}{x}\right) = {} & \frac{\Gamma(\gamma)\Gamma(\gamma-\alpha-\beta)}{\Gamma(\gamma-\alpha)\Gamma(\gamma-\beta)} \\ & \times x^{\alpha}{}_2F_1(\alpha, \alpha-\gamma+1; \alpha+\beta-\gamma+1; 1-x) \\ & + \frac{\Gamma(\gamma)\Gamma(\alpha+\beta-\gamma)}{\Gamma(\alpha)\Gamma(\beta)}x^{\beta}(x-1)^{\gamma-\alpha-\beta} \\ & \times {}_2F_1(\gamma-\alpha, 1-\alpha; \gamma-\alpha-\beta+1; 1-x), \qquad (9.3)\end{aligned}$$

where $1 < x < 2$ and $1 > \gamma > \alpha + \beta$.

These relations are typical of a larger number which exist between the solutions of the hypergeometric equation (8.1). If we change the independent variable in this equation to any one of

$$1-x, \frac{1}{x}, \frac{1}{1-x}, \frac{x-1}{x}, \frac{x}{x-1}$$

the equation transforms to one of the same type (but, of course, with different parameters). The equation (8.1) therefore has twelve solutions of the types (8.5) and (8.6) – two for each independent variable – each convergent within the unit circle. Any one of these can be expressed in terms of two fundamental solutions. In addition twelve more solutions of the kinds

$$(1-x)^{\gamma-\alpha-\beta}\,{}_2F_1(\gamma-\alpha, \gamma-\beta; \gamma; x),$$
$$x^{1-\gamma}(1-x)^{\gamma-\alpha-\beta}\,{}_2F_1(1-\alpha, 1-\beta; 2-\gamma; x)$$

can be derived. The relations between these twenty-four solutions of the hypergeometric equation are of the types (9.2) and (9.3); for a full discussion of them the reader is referred to MacRobert (1954), pp. 298–301.

§10. Relations of contiguity

Certain simple relations exist between hypergeometric functions whose parameters differ by ± 1. For example if the parameters α and β remain fixed and γ is varied we can prove that

$$\begin{aligned}\gamma\{\gamma-1-(2\gamma-1-\alpha-\beta)x\}\,{}_2F_1(\alpha, \beta; \gamma; x)&\\ +(\gamma-\alpha)(\gamma-\beta)x\,{}_2F_1(\alpha, \beta; \gamma+1; x)&\\ -\gamma(\gamma-1)(1-x)\,{}_2F_1(\alpha, \beta; \gamma-1; x)&=0. \quad (10.1)\end{aligned}$$

The proof follows from the definition (6.3). For the coefficient of x^n in the expansion of the function on the left of (10.1) is

$$\begin{aligned}&\gamma(\gamma-1)\frac{(\alpha)_n(\beta)_n}{(\gamma)_n n!}-(2\gamma-\alpha-\beta)\frac{(\alpha)_{n-1}(\beta)_{n-1}}{(\gamma)_{n-1}(n-1)!}\\ &\quad+(\gamma-\alpha)(\gamma-\beta)\frac{(\alpha)_{n-1}(\beta)_{n-1}}{(\gamma+1)_{n-1}(n-1)!}-\gamma(\gamma-1)\frac{(\alpha)_n(\beta)_n}{(\gamma-1)_n n!}\\ &\quad+\gamma(\gamma-1)\frac{(\alpha)_{n-1}(\beta)_{n-1}}{(\gamma-1)_{n-1}(n-1)!}\end{aligned}$$

and it is not difficult to show that this is zero.

In another kind β and γ are kept constant and α is varied. One such is

$$\{\gamma-\alpha-\beta+(\beta-\alpha)(1-x)\}\,{}_2F_1(\alpha,\beta;\gamma;x)$$
$$+\alpha(1-x)\,{}_2F_1(\alpha+1,\beta;\gamma;x)-(\gamma-\alpha)\,{}_2F_1(\alpha-1,\beta;\gamma;x)=0, \tag{10.2}$$

the proof of which is similarly direct.

In the third type of relation γ is kept constant and α and β vary. One of the simplest among these relations is

$$(\alpha-\beta)\,{}_2F_1(\alpha,\beta;\gamma;x)=\alpha\,{}_2F_1(\alpha+1,\beta;\gamma;x)$$
$$-\beta\,{}_2F_1(\alpha,\beta+1;\gamma;x) \tag{10.3}$$

The proof of these relations is left to the reader; further examples are given below (cf. Problems 2.3 and 2.4).

§11. The confluent hypergeometric function

If we replace x by x/β in equation (8.1) we see that the hypergeometric function

$${}_2F_1(\alpha,\beta;\gamma;x/\beta)$$

is a solution of the differential equation

$$x\left(1-\frac{x}{\beta}\right)\frac{d^2y}{dx^2}+\left\{\gamma-\left(1+\frac{\alpha+1}{\beta}\right)x\right\}\frac{dy}{dx}-\alpha y=0$$

so that letting $\beta\to\infty$ we see that the function

$$\lim_{\beta\to\infty}{}_2F_1(\alpha,\beta;\gamma;x/\beta) \tag{11.1}$$

is a solution of the differential equation

$$x\frac{d^2y}{dx^2}+(\gamma-x)\frac{dy}{dx}-\alpha y=0. \tag{11.2}$$

From the definition of $(\beta)_r$ we see that

$$\lim_{\beta\to\infty}\frac{(\beta)_r}{\beta^r}=1$$

so that the function (11.1) is the series

$$\sum_{r=0}^{\infty}\frac{(\alpha)_r}{(\gamma)_r}\frac{x^r}{r!}, \tag{11.3}$$

and this series we denote by the symbol ${}_1F_1(\alpha;\gamma;x)$. This function is called a **confluent hypergeometric function,** and the equation (11.2) is the **confluent hypergeometric equation.**

Equations of the type (11.2) occur in mathematical physics in the discussion of boundary value problems in potential theory, and in the theory of atomic collisions (see Problems 2.12, 2.13).

It is readily verified that the point $x=0$ is a regular point of the differential equation (11.2) and that, in the notation of §2.2, $p_0=\gamma$ and $q_0=0$. The indical equation is therefore

$$\rho(\rho+\gamma-1)=0$$

with roots $\rho=0$ and $\rho=1-\gamma$.

Corresponding to the root $\rho=0$ there is a solution of the form

$$y_1=\sum_{r=0}^{\infty} c_r x^r;$$

substituting this solution in equation (11.2) and equating to zero the coefficient of x^r we find that

$$c_{r+1}=\frac{(\alpha+r)c_r}{(\gamma+r)(r+1)}.$$

Putting $c_0=1$ we see that

$$c_r=\frac{(\alpha)_r}{(\gamma)_r}\frac{1}{r!},$$

and if γ is neither zero nor a negative integer the solution is

$$y_1(x)={}_1F_1(\alpha;\gamma;x). \tag{11.4}$$

Similarly, the root $\rho=1-\gamma$ leads, if $1-\gamma$ is neither zero nor a positive integer, to a solution of the type

$$y_2(x)=x^{1-\gamma}\sum_{r=0}^{\infty} c_r x^r.$$

If we write

$$y_2(x)=x^{1-\gamma}u(x),$$

and substitute in (11.2) we find that $u(x)$ satisfies the equation

$$x\frac{d^2u}{dx^2}+(2-\gamma-x)\frac{du}{dx}-(\alpha-\gamma+1)u=0,$$

which is the same as equation (11.2) with γ replaced by $2-\gamma$ and

α replaced by $\alpha-\gamma+1$. We know from equation (11.4) that the solution of this equation which has value unity when $x=0$ is $u={}_1F_1(\alpha-\gamma+1; 2-\gamma; x)$ so that

$$y_2(x)=x^{1-\gamma}\,{}_1F_1(\alpha-\gamma+1; 2-\gamma; x). \tag{11.5}$$

Thus if γ is neither 0 nor an integer the general solution of equation (11.2) is

$$y(x)=A\,{}_1F_1(\alpha;\gamma;x)+Bx^{1-\gamma}\,{}_1F_1(\alpha-\gamma+1; 2-\gamma; x), \tag{11.6}$$

where A and B are arbitrary constants.

In the exceptional case $\gamma=1$ we have

$$y_1(x)={}_1F_1(\alpha; 1; x), \tag{11.7}$$

obtained simply by putting $\gamma=1$ in equation (11.4). For the second solution we write

$$y_2(x)=y_1(x)\log x+\sum_{r=1}^{\infty} c_r x^r. \tag{11.8}$$

Substituting this expression in equation (11.2) we find that the unknown coefficients c_r must be such that

$$\frac{dy_1}{dx}-y_1+\sum_{r=1}^{\infty}\{(r+1)^2c_{r+1}-rc_r\}x^r+c_1=0.$$

Inserting the value of $y_1(x)$ from equation (11.7) we see that these coefficients are determined by the recurrence relation

$$c_1=1-\alpha,\ (r+1)^2c_{r+1}-rc_r=(1-\alpha)\frac{(\alpha)_r}{r!\,(r+1)!}. \tag{11.9}$$

The complete solution is therefore given by

$$y=Ay_1(x)+By_2(x),$$

where A and B are arbitrary constants and the functions $y_1(x)$, $y_2(x)$ are defined by equations (11.7), (11.8) and (11.9). The complete solution when γ is an integer may be found by a similar method.

If in equation (11.2) we put

$$y(x)=x^{-\frac{1}{2}\gamma}e^{\frac{1}{2}x}W(x) \tag{11.10}$$

we find that the function $W(x)$ satisfies the differential equation

$$\frac{d^2W}{dx^2}+\left\{-\frac{1}{4}+\frac{k}{x}+\frac{\frac{1}{4}-m^2}{x^2}\right\}W(x)=0, \tag{11.11}$$

where we have written k for $\frac{1}{2}\gamma-\alpha$ and m for $(\frac{1}{2}-\frac{1}{2}\gamma)$. The solutions of this equation are known as **Whittaker's confluent hypergeometric functions.**

If $2m$ is neither 1 nor an integer the solutions of the confluent hypergeometric equation corresponding to equation (11.11) are given by equation (11.6) with $\gamma=1+2m$ and $\alpha=\frac{1}{2}-k+m$. Thus the solutions of equation (11.11) are the Whittaker functions

$$M_{k,m}(x)=x^{\frac{1}{2}+m}e^{-\frac{1}{2}x}\,{}_1F_1(\tfrac{1}{2}-k+m;1+2m;x), \quad (11.12a)$$

$$M_{k,-m}(x)=x^{\frac{1}{2}-m}e^{-\frac{1}{2}x}\,{}_1F_1(\tfrac{1}{2}-k-m;1-2m;x). \quad (11.12b)$$

Several of the properties of ${}_2F_1$ functions have analogues for the ${}_1F_1$ functions. Corresponding to equation (7.1) there is the integral formula

$${}_1F_1(\alpha;\gamma;x)=\frac{1}{B(\alpha,\gamma-\alpha)}\int_0^1 (1-t)^{\gamma-\alpha-1}t^{\alpha-1}e^{xt}\,\mathrm{d}t, \quad (11.13)$$

from which Kummer's relation

$${}_1F_1(\alpha;\gamma;x)=e^x\,{}_1F_1(\gamma-\alpha;\gamma;-x) \quad (11.14)$$

may be obtained by a simple change of variable. The analogue of equation (6.5) is

$$\frac{\mathrm{d}}{\mathrm{d}x}\{{}_1F_1(\alpha;\gamma;x)\}=\frac{a}{\gamma}\,{}_1F_1(\alpha+1;\gamma+1;x), \quad (11.15)$$

while corresponding to the contiguity relations of §10 we have relations of the type

$$\alpha\,{}_1F_1(\alpha+1;\gamma+1;x)+(\gamma-\alpha)\,{}_1F_1(\alpha;\gamma+1;x) \\ -\gamma\,{}_1F_1(\alpha;\gamma;x)=0, \quad (11.16)$$

$$(x+\alpha)\,{}_1F_1(\alpha+1;\gamma+1;x)+(\gamma-\alpha)\,{}_1F_1(\alpha;\gamma+1;x) \\ -\gamma\,{}_1F_1(\alpha+1;\gamma;x)=0, \quad (11.17)$$

$$\alpha\,{}_1F_1(\alpha+1;\gamma;x)+(\gamma-2\alpha-x)\,{}_1F_1(\alpha;\gamma;x) \\ +(\alpha-\gamma)\,{}_1F_1(\alpha-1;\gamma;x)=0, \quad (11.18)$$

$$(\alpha-\gamma)x\,{}_1F_1(\alpha;\gamma+1;x)+\gamma(x+\gamma-1)\,{}_1F_1(\alpha;\gamma;x) \\ +\gamma(\gamma-1)\,{}_1F_1(\alpha;\gamma-1;x)=0. \quad (11.19)$$

§12. Generalised hypergeometric series

There are two ways by which we may approach the problem of generalising the idea of a hypergeometric function. We may think of such a function as being the solution of a linear differential equation which is an immediate generalisation of the equation (8.1) or we can define the function by a series which is analogous to the series (6.1).

At first sight it is difficult to see how the differential equation (8.1) can be generalised immediately, but if we introduce the operator

$$\vartheta = x\frac{\mathrm{d}}{\mathrm{d}x}$$

and notice that (8.1) is equivalent to

$$\{\vartheta(\vartheta+\gamma-1)-x(\vartheta+\alpha)(\vartheta+\beta)\}y=0, \tag{12.1}$$

an obvious generalisation is

$$\{\vartheta(\vartheta+\rho_1-1)\ldots(\vartheta+\rho_p-1)-x(\vartheta+\alpha_1)\ldots(\vartheta+\alpha_{p+1})\}y=0, \tag{12.2}$$

where $\alpha_1, \alpha_2, \ldots, \alpha_{p+1}, \rho_1, \ldots, \rho_p$ are constants. Furthermore it is readily shown that this equation is satisfied by the series

$$\sum_{n=0}^{\infty} \frac{(\alpha_1)_n(\alpha_2)_n\ldots(\alpha_{p+1})_n}{(\rho_1)_n(\rho_2)_n\ldots(\rho_p)_n}\frac{x^n}{n!}, \tag{12.3}$$

which is, itself, a generalisation of the series (6.1). Such a series is called a **generalised hypergeometric series** and is denoted by the symbol

$${}_{p+1}F_p(\alpha_1, \ldots, \alpha_{p+1}; \rho_1, \ldots, \rho_p; x).$$

It is left as an exercise to the reader to show that, if no two of the numbers $1, \rho_1, \rho_2, \ldots, \rho_p$ differ by an integer (or zero), the other p linearly independent solutions of equation (12.2) are

$$x^{1-\rho_i}\,{}_{p+1}E_p(1+\alpha_1-\rho_i, \ldots, 1+\alpha_{p+1}-\rho_i; 2-\rho_i,$$
$$1+\rho_1-\rho_2, \ldots, 1+\rho_p-\rho_i; x), \qquad i=1, 2, \ldots, n.$$

As it stands (12.3) is a generalisation of the series (6.1) but it is not sufficiently wide to cover a simple series of the type (11.3). To cover such cases we generalise, not the differential equation, but the series defining the function. The generalisation of (6.1) which includes (12.3) is the series

$$\sum_{n=0}^{\infty} \frac{(\alpha_1)_n(\alpha_2)_n \ldots (\alpha_p)_n}{(\rho_1)_n(\rho_2)_n \ldots (\rho_q)_n} \frac{x^n}{n!}, \tag{12.4}$$

which we denote by the symbol

$${}_pF_q(\alpha_1, \ldots, \alpha_p; \rho_1, \ldots, \rho_q; x),$$

or, if we wish particularly to throw into relief the difference between the numerator and the denominator parameters, by the symbol

$${}_pF_q\left[\begin{matrix} \alpha_1, \ldots, \alpha_p; x \\ \rho_1, \ldots, \rho_q; \end{matrix}\right].$$

The suffix p in front of the F denotes that there are p numerator parameters $\alpha_1, \ldots, \alpha_p$; similarly the suffix q indicates the number of denominator parameters.

Generalised hypergeometric series do not usually arise in mathematical physics because we have to solve equations of the type (12.2). Their use is more indirect. Such series occur normally only in the evaluation of integrals involving special functions. In certain cases these series reduce to series of the type

$${}_pF_q\left[\begin{matrix} \alpha_1, \ldots, \alpha_p; 1 \\ \rho_1, \ldots, \rho_q; \end{matrix}\right]$$

which have unit argument. For this reason it is desirable to have information about sums of this type. An account of the theory of such sums is given in Slater (1966). Here we shall consider only one such calculation because it illustrates the use of the theorems of Gauss and Kummer proved above (equations (7.2) and (7.3) respectively). Other results of this kind are given in Problems 2.18 and 2.20.

By expanding the ${}_3F_2$ series involved we see that

$$\begin{aligned} S &\equiv \frac{\Gamma(\alpha)\Gamma(\beta)\Gamma(\gamma)}{\Gamma(1+\alpha-\beta)\Gamma(1+\alpha-\gamma)}\,{}_3F_2\left[\begin{matrix} \alpha, & \beta, & \gamma; & 1 \\ 1+\alpha-\beta, & & 1+\alpha-\gamma; & \end{matrix}\right] \\ &= \sum_{n=0}^{\infty} \frac{\Gamma(\alpha+n)\Gamma(\beta+n)\Gamma(\gamma+n)}{n!\,\Gamma(1+\alpha-\beta+n)\Gamma(1+\alpha-\gamma+n)} \\ &= \sum_{n=0}^{\infty} \frac{\Gamma(\alpha+n)\Gamma(\beta+n)\Gamma(\gamma+n)}{n!\,\Gamma(1+\alpha+2n)\Gamma(1+\alpha-\beta-\gamma)} \\ &\quad \times \left\{\frac{\Gamma(1+\alpha+2n)\Gamma(1+\alpha-\beta-\gamma)}{\Gamma(1+\alpha-\beta+n)\Gamma(1+\alpha-\gamma+n)}\right\}. \end{aligned}$$

Now by Gauss's theorem (7.2) the expression inside the curly brackets is equal to ${}_2F_1(\beta+n, \gamma+n; 1+\alpha+2n; 1)$ which may be written

$$\sum_{m=0}^{\infty} \frac{\Gamma(\beta+n+m)\Gamma(\gamma+n+m)\Gamma(1+\alpha+2n)}{\Gamma(\beta+n)\Gamma(\gamma+n)\Gamma(1+\alpha+2n+m)m!};$$

whence we find that

$$S = \sum_{n=0}^{\infty} \sum_{m=0}^{\infty} \frac{\Gamma(\alpha+n)\Gamma(\beta+n+m)\Gamma(\gamma+n+m)}{\Gamma(1+\alpha+2n+m)\Gamma(1+\alpha-\beta-\gamma)n!m!}.$$

Interchanging the order of summation and putting $p = n+m$ we see that

$$S = \sum_{p=0}^{\infty} \frac{\Gamma(\beta+p)\Gamma(\gamma+p)}{\Gamma(1+\alpha-\beta-\gamma)} \sum_{n=0}^{p} \frac{\Gamma(\alpha+n)}{n!\,(p-n)!\Gamma(1+\alpha+n+p)}.$$

Now by Problem 1.27(iii)

$$\frac{1}{(p-n)!} = (-1)^n \frac{(-p)_n}{p!},$$

so that the inner sum is equal to

$$\frac{\Gamma(\alpha)}{p!\Gamma(1+\alpha+p)} {}_2F_1\left[\begin{matrix}\alpha, -p; -1 \\ 1+\alpha+p\end{matrix}\right],$$

which by Kummer's theorem (7.3) is equal to

$$\frac{\Gamma(\alpha)\Gamma(1+\frac{1}{2}\alpha)}{p!\Gamma(1+\alpha)\Gamma(1+\frac{1}{2}\alpha+p)}.$$

Therefore

$$S = \sum_{p=0} \frac{\Gamma(\alpha)\Gamma(\beta+p)\Gamma(\gamma+p)\Gamma(1+\frac{1}{2}\alpha)}{p!(1+\alpha-\beta-\gamma)\Gamma(1+\alpha)\Gamma(1+\frac{1}{2}\alpha+p)}$$
$$= \frac{\Gamma(\alpha)\Gamma(\beta)\Gamma(\gamma)}{\Gamma(1+\alpha)\Gamma(1+\alpha-\beta-\gamma)} {}_2F_1\left[\begin{matrix}\beta, \quad \gamma; \quad 1 \\ 1+\frac{1}{2}\alpha;\end{matrix}\right].$$

This ${}_2F_1$ series with unit argument can be summed by Gauss's formula (7.2) and the expression for S found. It then follows that

$${}_3F_2\left[\begin{matrix}\alpha, \quad \beta, \quad \gamma; \quad 1 \\ 1+\alpha-\beta, 1+\alpha-\gamma;\end{matrix}\right]$$
$$= \frac{\Gamma(1+\frac{1}{2}\alpha)\Gamma(1+\frac{1}{2}\alpha-\beta-\gamma)\Gamma(1+\alpha-\beta)\Gamma(1+\alpha-\gamma)}{\Gamma(1+\alpha)\Gamma(1+\alpha-\beta-\gamma)\Gamma(1+\frac{1}{2}\alpha-\beta)\Gamma(1+\frac{1}{2}\alpha-\gamma)},$$

a result which is known as **Dixon's theorem.**

Problems II

2.1 Show that

(i) ${}_2F_1(\alpha, \beta; \beta; z) = (1-z)^{-\alpha}$;

(ii) ${}_2F_1(\frac{1}{2}\alpha, \frac{1}{2}\alpha + \frac{1}{2}; \frac{1}{2}; z) = \frac{1}{2}\{(1-z)^{-\alpha} + (1+z)^{-\alpha}\}$;

(iii) ${}_2F_1(\frac{1}{2}\alpha + \frac{1}{2}, \frac{1}{2}\alpha + 1; \frac{3}{2}; z^2) = \dfrac{1}{2\alpha z}\{(1-z)^{-\alpha} - (1+z)^{-\alpha}\}$;

(iv) ${}_2F_1(1, 1; 2; z) = -\dfrac{1}{z}\log(1-z)$;

(v) ${}_2F_1(\frac{1}{2}, 1; \frac{3}{2}; z^2) = \dfrac{1}{2z}\log\dfrac{1+z}{1-z}$;

(vi) ${}_2F_1(\frac{1}{2}, \frac{1}{2}; \frac{3}{2}; z^2) = \dfrac{\sin^{-1} z}{z}$;

(vii) ${}_2F_1(\frac{1}{2}, 1; \frac{3}{2}; -z^2) = \dfrac{\tan^{-1} z}{z}$;

(viii) ${}_2F_1(\frac{1}{2}, \frac{1}{2}; 1; k^2) = \dfrac{2}{\pi}K(k)$;

(ix) ${}_2F_1(-\frac{1}{2}, \frac{1}{2}; 1; k^2) = \dfrac{2}{\pi}E(k)$.

2.2 By transforming the equation $y'' + n^2 y = 0$ to hypergeometric form by the substitution $\xi = \sin^2 z$, prove that, if $-\frac{1}{2}\pi \leqslant z \leqslant \frac{1}{2}\pi$,

(i) $\cos(nz) = {}_2F_1(\frac{1}{2}n, -\frac{1}{2}n; \frac{1}{2}; \sin^2 z)$;

(ii) $\sin(nz) = n \sin z \; {}_2F_1(\frac{1}{2} - \frac{1}{2}n, \frac{1}{2} + \frac{1}{2}n; \frac{3}{2}; \sin^2 z)$;

and that, if $0 \leqslant z \leqslant \pi$,

(iii) $\cos(nz) = \cos(\frac{1}{2}n\pi)\, {}_2F_1(\frac{1}{2}n, -\frac{1}{2}n; \frac{1}{2}; \cos^2 z)$
$\qquad + n \sin(\frac{1}{2}n\pi)\cos(z)\, {}_2F_1(\frac{1}{2} - \frac{1}{2}n, \frac{1}{2} + \frac{1}{2}n; \frac{3}{2}; \cos^2 z)$;

(iv) $\sin(nz) = \sin(\frac{1}{2}n\pi)\, {}_2F_1(\frac{1}{2}n, -\frac{1}{2}n; \frac{1}{2}; \cos^2 z)$
$\qquad - n \cos(\frac{1}{2}n\pi)\cos(z)\, {}_2F_1(\frac{1}{2} - \frac{1}{2}n, \frac{1}{2} + \frac{1}{2}n; \frac{3}{2}; \cos^2 z)$;

2.3 Prove the relations:

(i) $(\alpha - \beta)(1-x)\, {}_2F_1(\alpha, \beta; \gamma; x)$
$\qquad = (\gamma - \beta)\, {}_2F_1(\alpha, \beta - 1; \gamma; x) - (\gamma - \alpha)\, {}_2F_1(\alpha - 1, \beta; \gamma; x)$;

(ii) $(\gamma-\beta-1)\,{}_2F_1(\alpha, \beta; \gamma; x)$

$$=(\gamma-\alpha-\beta-1)\,{}_2F_1(\alpha, \beta+1; \gamma; x)+\alpha(1-x)\,{}_2F_1(\alpha+1, \beta+1; \gamma; x);$$

$$=(\alpha-\beta-1)(1-x)\,{}_2F_1(\alpha, \beta+1; \gamma; x) +(\gamma-\alpha)\,{}_2F_1(\alpha-1, \beta+1; \gamma; x);$$

(iii) $(\gamma-\alpha-\beta)\,{}_2F_1(\alpha, \beta; \gamma; x)$

$$=(\gamma-\alpha)\,{}_2F_1(\alpha-1, \beta; \gamma; x)-\beta(1-x)\,{}_2F_1(\alpha, \beta+1; \gamma; x);$$

(iv) $\alpha\,{}_2F_1(\alpha+1; \beta; \gamma; x)-(\gamma-1)\,{}_2F_1(\alpha, \beta; \gamma-1; x)$

$$=(\alpha+1-\gamma)\,{}_2F_1(\alpha, \beta; \gamma; x);$$

(v) $(1-x)\,{}_2F_1(\alpha, \beta; \gamma; x)-{}_2F_1(\alpha-1, \beta-1; \gamma; x)$

$$=\frac{\alpha+\beta-\gamma-1}{\gamma}\,x\,{}_2F_1(\alpha, \beta; \gamma+1; x);$$

(vi) $\dfrac{(1-\beta)x}{\gamma}\,{}_2F_1(\alpha, \beta; \gamma+1; x)$

$$={}_2F_1(\alpha-1, \beta-1; \gamma; x)-{}_2F_1(\alpha, \beta-1; \gamma; x);$$

(viii) $(1-x)\,{}_2F_1(\alpha, \beta; \gamma; x)$

$$={}_2F_1(\alpha, \beta-1; \gamma; x)+\frac{(\alpha-\gamma)x}{\gamma}\,{}_2F_1(\alpha, \beta; \gamma+1; x).$$

2.4 Prove that

(i) ${}_2F_1(\alpha, \beta+1; \gamma+1; z)-{}_2F_1(\alpha, \beta; \gamma; z)$

$$=\frac{\alpha(\gamma-\beta)}{\gamma(\gamma+1)}\,z\,{}_2F_1(\alpha+1, \beta+1; \gamma+2; z);$$

(ii) ${}_2F_1(\alpha, \beta; \gamma; z)={}_2F_1(\alpha+1, \beta-1; \gamma; z)$

$$+\frac{\alpha-\beta+1}{\gamma}\,z\,{}_2F_1(\alpha+1, \beta; \gamma+1; z).$$

Deduce a simple expression for the hypergeometric series ${}_2F_1(\alpha, \beta; \beta-1; z)$.

2.5 Prove that n is a positive integer,

$${}_2F_1(-n, \alpha+n; \gamma; x)=\frac{x^{1-\gamma}(1-x)^{\gamma-\alpha}\Gamma(\gamma)}{\Gamma(\gamma+n)}\frac{d^n}{dx^n}\{x^{\gamma+n-1}(1-x)^{\alpha-\gamma+n}\},$$

and deduce that

$${}_2F_1\left(-n, \alpha+n; \tfrac{1}{2}\alpha+\tfrac{1}{2}; \frac{1-\mu}{2}\right)=\frac{(\mu^2-1)^{\frac{1}{2}-\frac{1}{2}\alpha}\Gamma(\frac{1}{2}\alpha+\frac{1}{2})}{2^n\Gamma(\frac{1}{2}\alpha+\frac{1}{2}+n)}\frac{d^n}{d\mu^n}(\mu^2-1)^{n+\frac{1}{2}\alpha-\frac{1}{2}}.$$

2.6 Given that n is a positive integer and $|x|>1$, prove that

$$ {}_2F_1\left(\frac{n+1}{2}, \frac{n+2}{2}; 1; -\frac{1}{x^2}\right) = \frac{(-1)^n x^{n+1}}{n!} \frac{d^n}{dx^n} \left\{\frac{1}{\sqrt{(x^2+1)}}\right\}. $$

2.7 Establish the following formulae:

(i) ${}_2F_1(\alpha; \beta; \alpha+\beta+\rho; x) \times {}_2F_1(\gamma; \delta; \gamma+\delta-\rho; x)$

$$ = {}_2F_1(\alpha+\rho, \beta+\rho; \alpha+\beta+\rho; x) \times {}_2F_1(\gamma-\rho, \delta-\rho; \gamma+\delta-\rho; x); $$

(ii) $D_x\{x^\alpha\, {}_2F_1(\alpha, \beta; \gamma; kx)\} = \alpha x^{\alpha-1}\, {}_2F_1(\alpha+1; \beta; \gamma; kx)$, where $\alpha \neq 0$;

(iii) $B(\lambda, \gamma-\lambda)\, {}_2F_1(\alpha, \beta; \gamma; x)$

$$ = \int_0^1 t^{\lambda-1}(1-t)^{\gamma-\lambda-1}\, {}_2F_1(\alpha, \beta; \lambda; xt)\, dt, $$

where $|x|<1, \lambda>0, \gamma-\lambda>0$.

2.8 Prove that if $\beta>0$,

$$ {}_2F_1(\alpha, \beta; 2\beta; z) = \frac{(1-\frac{1}{2}z)^{-\alpha}}{2^{2\beta-1}B(\beta, \beta)} \int_0^{\frac{1}{2}\pi} (\sin\phi)^{2\beta-1} \left[\begin{matrix} \{1+\zeta\cos\phi\}^{-\alpha} \\ +\{1-\zeta\cos\phi\}^{-\alpha} \end{matrix}\right] d\phi, $$

where $\zeta = z/(2-z)$.

Deduce that

$$ {}_2F_1(\alpha, \beta; 2\beta; z) = (1-\tfrac{1}{2}z)^{-\alpha}\, {}_2F_1(\tfrac{1}{2}\alpha, \tfrac{1}{2}\alpha+\tfrac{1}{2}; \beta+\tfrac{1}{2}; \zeta^2). $$

2.9 Show that the equation

$$ (1-x^2)\frac{d^2y}{dx^2} - (\alpha+\beta+1)\frac{dy}{dx} - \alpha\beta y = 0, $$

has a solution

$$ {}_2F_1\left(\alpha, \beta; \tfrac{1}{2}\alpha+\tfrac{1}{2}\beta+\tfrac{1}{2}; \frac{1-x}{2}\right) $$

in the segment $-1<x<3$. Write down the solution in the case in which $\alpha+\beta$ is not an odd integer.

Deduce the general solution of the equation

$$ \frac{d^2y}{d\theta^2} + (\alpha+\beta)\cot\theta \frac{dy}{d\theta} - \alpha\beta y = 0 $$

on the same assumption concerning α and β.

2.10 Schrödinger's equation for the rotation of a symmetrical-top molecule is

$$\frac{1}{\sin\theta}\frac{\partial}{\partial\theta}\left(\sin\theta\frac{\partial\psi}{\partial\theta}\right)+\frac{1}{\sin^2\theta}\frac{\partial^2\psi}{\partial\theta^2}$$
$$+\left(\cot^2\theta+\frac{A}{C}\right)\frac{\partial^2\psi}{\partial\chi^2}-\frac{2\cos\theta}{\sin^2\theta}\frac{\partial^2\psi}{\partial\chi\,\partial\phi}+\frac{8\pi^2 AW}{h^2}\psi=0,$$

where A, C, W, h are constants. Show that it possesses solutions of the form

$$\psi=e^{im\phi-in\chi}(1-x)^{\frac{1}{2}(n-m)}x^{\frac{1}{2}(n-m)}\,{}_2F_1(\alpha,\beta;\gamma;x),$$

where $n\geqslant m$, $x=\frac{1}{2}(1-\cos\theta)$, $\gamma=n-m+1$, and α, β are the roots of the equation

$$z^2-(2n+1)z+\frac{A}{C}n^2+n-\frac{8\pi^2 AW}{h^2}=0.$$

2.11 Prove that:

(i) ${}_1F_1(\alpha;\alpha;x)=e^x;$

(ii) ${}_1F_1(\alpha+1;\alpha;x)=\left(1+\dfrac{x}{\alpha}\right)e^x;$

(iii) ${}_1F_1(\frac{1}{2};\frac{3}{2};-x^2)=\dfrac{\sqrt{\pi}}{2x}\operatorname{erf}(x);$

(iv) ${}_1F_1(\alpha+1;\gamma;x)-{}_1F_1(\alpha;\gamma;x)=\dfrac{x}{\gamma}\,{}_1F_1(\alpha+1;\gamma+1;x);$

(v) ${}_1F_1(-\frac{1}{2};\frac{1}{2};-x^2)=e^{-x^2}-\sqrt{\pi}\,x\operatorname{erf}(x);$

(vi) $x^n\,{}_1F_1(n;n+1;-x)=n\displaystyle\int_0^x t^{n-1}e^{-t}\,dt.$

2.12 Prove that the equation

$$\frac{\partial^2 V}{\partial x^2}=\frac{1}{k}\frac{\partial V}{\partial t}$$

possesses solutions of the type

$$V=Ct^m\,{}_1F_1\left(-m;\tfrac{1}{2};-\frac{x^2}{4kt}\right),$$

where m and C are constants.

2.13 Show that the Schrödinger equation

$$\nabla^2\psi + \left(k^2 - \frac{\beta}{r}\right)\psi = 0$$

possesses a solution of the form

$$e^{ikz}\ {}_1F_1\left(-\frac{i\beta}{2k}; 1; ikr - ikz\right).$$

2.14 The Schrödinger equation governing the radial wave functions for positive energy states in a Coulomb field is

$$\frac{1}{r}\frac{\mathrm{d}}{\mathrm{d}r}\left(r\frac{\mathrm{d}L}{\mathrm{d}r}\right) + \left[\frac{8\pi^2 m}{h^2}\left(W - \frac{zz'e^2}{r}\right) - \frac{n(n+1)}{r^2}\right]L = 0.$$

Show that it possesses a solution

$$L = r^n e^{ikr}\ {}_1F_1(i\alpha + n + 1; 2n+2; -2ikr),$$

where $k^2 = 8\pi^2 mW/h^2$, $\alpha = 4\pi^2 mzz'e^2/k$.

2.15 Show that the equation

$$\frac{\mathrm{d}^2y}{\mathrm{d}x^2} + \frac{1}{x}\frac{\mathrm{d}y}{\mathrm{d}x} + \left\{m^2 - \frac{2m\beta}{x} - \frac{n^2}{x^2}\right\}y = 0$$

possesses a solution

$$y = x^{\frac{1}{2}n} e^{-\frac{1}{2}x}\ {}_1F_1(\tfrac{1}{2}n + \tfrac{1}{2} - i\beta; 2imx),$$

and hence that a solution of equation (1.9c) is

$$R = \rho^{\frac{1}{2}n} e^{-\frac{1}{2}\rho} F(\tfrac{1}{2}n + \tfrac{1}{2}; n+1; 2im\rho).$$

2.16 Show that

(i) $$\int_0^1 x^{l-1}(1-x)^{m-1}\ {}_pF_q\left[\begin{matrix}\alpha_1, \ldots, \alpha_p, \kappa x\\ \beta_1, \ldots, \beta_q;\end{matrix}\right]\mathrm{d}x$$

$$= B(1, m)_{p+1}F_{q+1}\left[\begin{matrix}\alpha_1, \ldots, \alpha_p, l; \kappa\\ \beta_1, \ldots, \beta_q, l+m;\end{matrix}\right];$$

(ii) $$\int_0^1 x^{l-1}(1-x)^{m-1}\ {}_pF_q\left[\begin{matrix}\alpha_1, \ldots, \alpha_p; \dfrac{1-x}{2}\\ \beta_1, \ldots, \beta_q;\end{matrix}\right]\mathrm{d}x$$

$$= B(1, m)_{p+1}F_{q+1}\left[\begin{matrix}\alpha_1, \ldots, \alpha_p, m; \frac{1}{2}\\ \beta_1, \ldots, \beta_q, l+m\end{matrix}\right];$$

(iii) $$\int_0^1 (1-x^2)^{m-1}\,{}_pF_q\left[\begin{matrix}\alpha_1,\ldots,\alpha_p; \dfrac{1-x}{2}\\ \beta_1,\ldots,\beta_q\end{matrix}\right]dx$$
$$=B(\tfrac{1}{2},m)\,{}_{p+1}F_{q+1}\left[\begin{matrix}\alpha_1,\ldots,\alpha_p, m; 1\\ \beta_1,\ldots,\beta_q, 2m;\end{matrix}\right].$$

2.17 Prove that

(i) $$\int_0^\infty {}_pF_q\left[\begin{matrix}\alpha_1,\ldots,\alpha_p; bx\\ \beta_1,\ldots,\beta_q;\end{matrix}\right]e^{-ax}x^{\mu-1}\,dx$$
$$=\frac{\Gamma(\mu)}{a^\mu}\,{}_{p+1}F_q\left[\begin{matrix}\alpha_1,\ldots,\alpha_p,\mu; b/a\\ \beta_1,\ldots,\beta_q;\end{matrix}\right];$$

(ii) $$\int_0^\infty {}_pF_q\left[\begin{matrix}\alpha_1,\ldots,\alpha_p; \pm b^2x^2\\ \beta_1,\ldots,\beta_q;\end{matrix}\right]e^{-ax}x^{\mu-1}\,dx$$
$$=\frac{\Gamma(\mu)}{a^\mu}\,{}_{p+2}F_q\left[\begin{matrix}\alpha_1,\ldots,\alpha_p,\tfrac{1}{2}\mu,\tfrac{1}{2}\mu+\tfrac{1}{2}; \pm 4b^2/a^2\\ \beta_1,\ldots,\beta_q;\end{matrix}\right];$$

(iii) $$\int_0^\infty {}_pF_q\left[\begin{matrix}\alpha_1,\ldots,\alpha_p; \pm b^2x^2\\ \beta_1,\ldots,\beta_q;\end{matrix}\right]e^{-p^2x^2}x^{\mu-1}\,dx$$
$$=\frac{\Gamma(\tfrac{1}{2}\mu)}{2p^\mu}\,{}_{p+1}F_q\left[\begin{matrix}\alpha_1,\ldots,\alpha_p,\tfrac{1}{2}\mu; \pm b^2/p^2\\ \beta_1,\ldots,\beta_q;\end{matrix}\right].$$

2.18 By equating coefficients of x in the relation

$$(1-x)^{\alpha-\beta-\gamma}\,{}_2F_1(\alpha,\beta;\gamma;x)={}_2F_1(\gamma-\alpha,\gamma-\beta;\gamma;x),$$

prove **Saalschutz's theorem**

$${}_3F_2\left[\begin{matrix}\alpha,\beta,-n; 1\\ \gamma, 1+\alpha+\beta-\gamma-n\end{matrix}\right]=\frac{(\gamma-\alpha)_n(\gamma-\beta)_n}{(\gamma)_n(\gamma-\alpha-\beta)_n}.$$

Hence prove that

$${}_2F_1\left[\begin{matrix}\alpha,\beta; x\\ 1+\alpha-\beta\end{matrix}\right]=(1-x)^{-\alpha}\,{}_2F_1\left[\begin{matrix}\tfrac{1}{2}\alpha,\tfrac{1}{2}+\tfrac{1}{2}\alpha-\beta; -\dfrac{4x}{(1-x)^2}\\ 1+\alpha-\beta;\end{matrix}\right]$$

if $|x|<3-2\sqrt{2}$.

2.19 Show that

$${}_2F_1(\alpha,\beta;1+\alpha-\beta;z)=(1-z)^{-\alpha}\,{}_2F_1(\tfrac{1}{2}\alpha,\tfrac{1}{2}+\tfrac{1}{2}\alpha-\beta;1+\alpha-\beta;\zeta)$$

where $\zeta=-4z(1-z)^{-2}$. Hence deduce the value of ${}_2F_1(\alpha,\beta;1+\alpha-\beta;-1)$ from Gauss's theorem.

2.20 Show that

$$ {}_3F_2\left[\begin{matrix}\alpha, \beta, \gamma; 1\\ \delta, \varepsilon;\end{matrix}\right] = \frac{\Gamma(\delta)\Gamma(\varepsilon)\Gamma(\sigma)}{\Gamma(\alpha)\Gamma(\sigma+\beta)\Gamma(\sigma+\gamma)}\, {}_3F_2\left[\begin{matrix}\delta-\alpha, \varepsilon-\alpha, \sigma; 1\\ \sigma+\beta, \sigma+\gamma\end{matrix}\right], $$

where $\sigma = \delta+\varepsilon-\alpha-\beta-\gamma$.

Hence, using Dixon's theorem, prove **Watson's theorem:**

$$ {}_3F_2\left[\begin{matrix}\alpha, \beta, \gamma; 1\\ \frac{1}{2}(1+\alpha+\beta), 2\gamma\end{matrix}\right] = \frac{\Gamma(\frac{1}{2})\Gamma(\frac{1}{2}+\gamma)\Gamma(\frac{1}{2}+\frac{1}{2}\alpha+\frac{1}{2}\beta)\Gamma(\frac{1}{2}-\frac{1}{2}\alpha-\frac{1}{2}\beta+\gamma)}{\Gamma(\frac{1}{2}+\frac{1}{2}\alpha)\Gamma(\frac{1}{2}+\frac{1}{2}\beta)\Gamma(\frac{1}{2}-\frac{1}{2}\alpha+\gamma)\Gamma(\frac{1}{2}-\frac{1}{2}\beta+\gamma)} $$

and, using Watson's theorem, deduce **Whipple's theorem** that, if $\alpha+\beta=1$, and $\varepsilon+\delta=2\gamma+1$,

$$ {}_3F_2\left[\begin{matrix}\alpha, \beta, \gamma; 1\\ \delta, \varepsilon;\end{matrix}\right] = \frac{\pi\Gamma(\delta)\Gamma(\varepsilon)}{2^{2\gamma-1}\Gamma(\frac{1}{2}\alpha+\frac{1}{2}\delta)\Gamma(\frac{1}{2}\alpha+\frac{1}{2}\varepsilon)\Gamma(\frac{1}{2}\beta+\frac{1}{2}\delta)\Gamma(\frac{1}{2}\beta+\frac{1}{2}\varepsilon)}. $$

3

Legendre functions

§13. Legendre polynomials

If A is a fixed point with coordinates (α, β, γ) and P is the variable point (x, y, z), then if we denote the distance AP by R, we have

$$R^2 = (x-\alpha)^2 + (y-\beta)^2 + (z-\gamma)^2.$$

Furthermore, we know from elementary considerations that

$$\psi = \frac{1}{R}$$

is the gravitational potential at the point P due to a unit mass situated at the point A, and that this must be a particular solution of Laplace's equation.

In some circumstances it is desirable to expand ψ in powers of r or r^{-1} where $r = (x^2+y^2+z^2)^{\frac{1}{2}}$ is the distance of P from O, the origin of coordinates. This expansion can be obtained by the use of Taylor's theorem for functions of three variables but it is much more suitable to introduce the angle θ between the directions OA, OP and write

$$R^2 = r^2 + a^2 - 2ar\cos\theta.$$

The expression for ψ then becomes

$$\psi = \frac{1}{\sqrt{(a^2 - 2ar\mu + r^2)}} \tag{13.1}$$

where μ denotes $\cos\theta$, and this can be expanded in powers of r/a when $r < a$ and in powers of a/r when $r > a$. If we denote by $P_n(\mu)$ the coefficient of h^n in the expansion of $(1-2\mu h+h^2)^{-\frac{1}{2}}$ in ascending powers of h, i.e. if

$$\frac{1}{\sqrt{(1-2\mu h+h^2)}} = \sum_{n=0}^{\infty} P_n(\mu)h^n, \tag{13.2}$$

then the potential function (13.1) can be expanded in the forms

$$\frac{1}{a}\sum_{n=0}^{\infty}\left(\frac{r}{a}\right)^n P_n(\mu), \qquad r<a; \tag{13.3a}$$

$$\frac{1}{r}\sum_{n=0}^{\infty}\left(\frac{a}{r}\right)^n P_n(\mu), \qquad r>a. \tag{13.3b}$$

It is clear from the definition (13.2) that the coefficients $P_n(\mu)$ are polynomials in μ. The first one or two can readily be calculated directly from the definition. By the binomial theorem we have

$$(1-2\mu h+h^2)^{-\frac{1}{2}}=1+(-\tfrac{1}{2})(-2\mu h+h^2)+\frac{(-\frac{1}{2})(-\frac{3}{2})}{2}(-2\mu h+h^2)^2+\ldots$$

$$=1+\mu h+\tfrac{1}{2}(3\mu^2-1)h^2+\tfrac{1}{2}(5\mu^3-3\mu)h^3+\ldots,$$

so that

$$P_0(\mu)=1, \qquad P_1(\mu)=\mu,$$
$$P_2(\mu)=\tfrac{1}{2}=\tfrac{1}{2}(3\mu^2-1), \qquad P_3(\mu)=\tfrac{1}{2}(5\mu^3-3\mu). \tag{13.4a}$$

We shall show that, in the general case, $P_n(\mu)$ is a polynomial in μ degree n; it is called the **Legendre polynomial of order *n*.**

The expression for the general polynomial $P_n(\mu)$ can be derived by the method employed to obtain the simple expressions (13.4a).

Expanding $(1-2\mu h+h^2)^{-\frac{1}{2}}$ by the binomial theorem we have

$$(1-2\mu h+h^2)^{-\frac{1}{2}}=\sum_{r=0}^{\infty}\frac{(\frac{1}{2})_r}{r!}(2\mu h-h^2)^r,$$

and the coefficient of h^n in this expansion is the coefficient of h^n in the expansion

$$\sum_{r=0}^{n}\frac{(\frac{1}{2})_r}{r!}(2\mu h-h^2)^r=\sum_{\rho=0}^{n}\frac{(\frac{1}{2})_{n-\rho}}{(n-\rho)!}(2\mu h-h^2)^{n-\rho}$$

$$=\sum_{\rho=0}^{n}(-1)^{\rho}\frac{(\frac{1}{2})_n}{(\frac{1}{2}-n)_\rho}\frac{(2\mu h-h^2)^{n-\rho}}{(n-\rho)!},$$

since by Problem 1.27(ii)

$$(-1)^{\rho}(\tfrac{1}{2})_{n-\rho}=\frac{(\frac{1}{2})_n}{(\frac{1}{2}-n)_\rho}.$$

Now the coefficient of h^n in the expansion of

$$\frac{(-1)^\rho}{(n-\rho)!}(2\mu h-h^2)^{n-\rho}$$

is

$$\frac{(2\mu)^{n-2\rho}}{\rho!(n-2\rho)!},$$

and, by the duplication formula for the gamma function,

$$\frac{n!}{(n-2\rho)!}2^{-2\rho}=\frac{\Gamma(\frac{1}{2}n+\frac{1}{2})}{\Gamma(\frac{1}{2}n+\frac{1}{2}-\rho)}\frac{\Gamma(\frac{1}{2}n+1)}{\Gamma(\frac{1}{2}n+1-\rho)}=(\tfrac{1}{2}-\tfrac{1}{2}n)_\rho(-\tfrac{1}{2}n)_\rho,$$

so that

$$P_n(\mu)=\frac{(\frac{1}{2})_n}{n!}(2\mu)^n\sum_{\rho=0}^{\infty}\frac{(\frac{1}{2}-\frac{1}{2}n)_\rho(-\frac{1}{2}n)_\rho}{\rho!(\frac{1}{2}-n)_\rho}\left(\frac{1}{\mu^2}\right)^\rho,$$

a result which may be written in the form

$$P_n(\mu)=\frac{(2\mu)^n(\frac{1}{2})_n}{n!}\,{}_2F_1\left(\tfrac{1}{2}-\tfrac{1}{2}n,-\tfrac{1}{2}n;\tfrac{1}{2}-n;\frac{1}{\mu^2}\right). \tag{13.4b}$$

Putting $\mu=1$ in equation (13.2) and equating coefficients of h^n we find that

$$P_n(1)=1 \tag{13.5a}$$

for all values of n. Similarly if we put $\mu=-1$ in (13.2) we derive the result

$$P_n(-1)=(-1)^n, \tag{13.5b}$$

which is a particular case of the result

$$P_n(-\mu)=(-1)^nP_n(\mu). \tag{13.6}$$

Equation (13.4) gives $P_n(\cos\theta)$ as a polynomial in $\cos\theta$ of degree n so that it should be possible to express $P_n(\cos\theta)$ in terms of cosines of multiples of θ. Instead of attempting to do this by substituting the appropriate expression for $\cos^r\theta$ in (13.4) we begin afresh with the definition (13.2). Writing $(1-2\cos\theta h+h^2)$ in the form $(1-e^{i\theta}h)(1-e^{-i\theta}h)$, we find that

$$\sum_{n=0}^{\infty}P_n(\cos\theta)h^n=(1-he^{i\theta})^{-\frac{1}{2}}(1-he^{-i\theta})^{-\frac{1}{2}}$$

$$=\sum_{r=0}^{\infty}\sum_{s=0}^{\infty}\frac{\Gamma(r+\frac{1}{2})\Gamma(s+\frac{1}{2})}{\Gamma(\frac{1}{2})\Gamma(\frac{1}{2})r!s!}h^{r+s}e^{i(r-s\theta)}.$$

Equating the coefficients of h^n we find that

$$P_n(\cos\theta)=\sum_{r=0}^{n}\frac{\Gamma(\frac{1}{2}+r)\Gamma(\frac{1}{2}+n-r)}{\Gamma(\frac{1}{2})\Gamma(\frac{1}{2})r!(n-r)!}e^{i(2r-n)\theta}.$$

Using the duplication formula, we see that

$$\frac{\Gamma(\frac{1}{2}+r)\Gamma(\frac{1}{2}+n-r)}{\Gamma(\frac{1}{2})\Gamma(\frac{1}{2})}=\frac{1}{2^{2n}}\frac{(2n-2r)!(2r)!}{r!(n-r)!},$$

so that

$$P_n(\cos\theta)=\frac{1}{2^{2n}}\sum_{r=0}^{n}\frac{(2n-2r)!(2r)!}{(r!)^2\{(n-r)!\}^2}e^{i(2r-n)\theta},$$

from which it follows immediately that

$$P_{2n}(\cos\theta)=\frac{\{(2n)!\}^2}{2^{4n}(n!)^4}+\frac{1}{2^{4n-1}}\sum_{r=0}^{n-1}\frac{(4n-2r)!(2r)!}{(r!)^2\{(2n-r)!\}^2}\cos(2n-2r)\theta, \tag{13.7}$$

and

$$P_{2n+1}(\cos\theta)=\frac{1}{2^{4n+1}}\sum_{r=0}^{n}\frac{(4n+2-2r)!(2r)!}{(r!)^2\{(2n+1-r)!\}^2}\cos(2n-2r+1)\theta. \tag{13.8}$$

From these last two equations we may derive a general result of some importance. We may write

$$P_n(\cos\theta)=\sum_{r=0}^{p}c_r\cos(n-2r)\theta, \tag{13.9}$$

where $p=\frac{1}{2}n$ or $\frac{1}{2}n-\frac{1}{2}$ according as n is even or odd. In particular

$$P_n(1)=\sum_{r=0}^{p}c_r,$$

so that from (13.5)

$$1=\sum_{r=0}^{p}c_r.$$

Now, from equation (13.9) we have

$$|P_n(\cos\theta)|\leqslant\sum_{r=0}^{p}c_r.$$

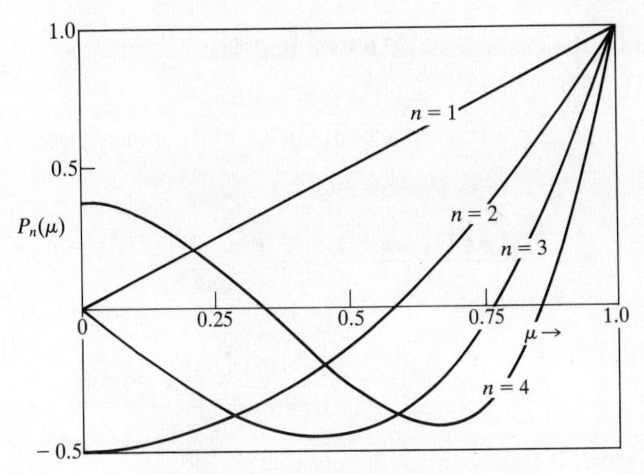

Fig. 5 Variation of $P_n(\mu)$ with μ

and therefore

$$|P_n(\cos\theta| \leqslant 1. \tag{13.10}$$

The variation of $P_n(\mu)$ with μ for a few values of n is shown in Fig. 5. Since, in most physical problems, the Legendre polynomial involved is usually $P_n(\cos\theta)$ we have shown in Fig. 6 the variation of this function with θ. Numerical values may be obtained from Zhurina and Karmazina (1966).

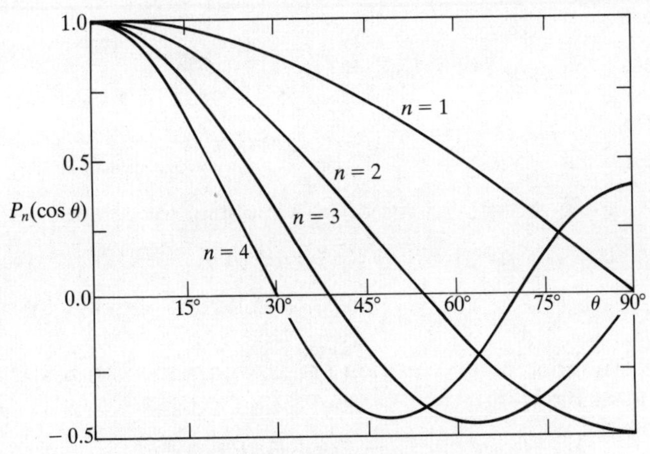

Fig. 6 Variation of $P_n(\cos\theta)$ with θ

§14. Recurrence relations for the Legendre polynomials

If we differentiate both sides of equation (13.2) with respect to h we have

$$\frac{\mu-h}{(1-2\mu h+h^2)^{\frac{3}{2}}}=\sum_{n=0}^{\infty} nh^{n-1}P_n(\mu), \qquad |\mu|<1,$$

which may be written in the form

$$(\mu-h)\sum_{n=0}^{\infty} h^n P_n(\mu)=(1-2\mu h+h^2)\sum_{n=0}^{\infty} nh^{n-1}P_n(\mu). \tag{14.1}$$

Equating coefficients of h^n we have

$$\mu P_n(\mu)-P_{n-1}(\mu)=(n+1)P_{n+1}(\mu)-2n\mu P_n(\mu)+(n-1)P_{n-1}(\mu),$$

which reduces to

$$(n+1)P_{n+1}(\mu)-(2n+1)\mu P_n(\mu)+nP_{n-1}(\mu)=0. \tag{14.2}$$

This relation has been proved to hold for $|\mu|<1$ but since the left-hand side is a polynomial in μ, it must hold for *all* values of μ.

On the other hand, if we differentiate both sides of equation (13.2) with respect to μ we obtain the relation

$$\frac{h}{(1-2\mu h+h^2)^{\frac{3}{2}}}=\sum_{n=0}^{\infty} h^n P_n'(\mu). \tag{14.3}$$

Combining equations (14.1) and (14.3), we have

$$(\mu-h)\sum_{n=0}^{\infty} h^n P_n'(\mu)=\sum_{n=0}^{\infty} nh^n P_n(\mu),$$

so that equating the coefficients of h^n we obtain the relation

$$\mu P_n'(\mu)-P_{n-1}'(\mu)=nP_n(\mu), \tag{14.4}$$

and since each side is a polynomial in μ this relation holds for all values of μ.

If now we differentiate equation (14.2) with respect to μ we obtain the relation

$$(n+1)P_{n+1}'(\mu)-(2n+1)P_n(\mu)-(2n+1)\mu P_n'(\mu)+nP_{n-1}'(\mu)=0. \tag{14.5}$$

Eliminating $P_n'(\mu)$ from (14.4) and (14.5) we see that

$$P'_{n+1}(\mu) - P'_{n-1}(\mu) = (2n+1)P_n(\mu). \tag{14.6}$$

Subtracting (14.4) from (14.6) we obtain the recurrence relation

$$P'_{n+1}(\mu) - \mu P'_n(\mu) = (n+1)P_n(\mu). \tag{14.7}$$

The differentiations with respect to h and μ under the summation sign is justified by the fact that the series on the right-hand side of equation (13.2) is uniformly convergent for all real or complex values of h and μ which satisfy the relation $|\mu| \leqslant 1$, $|h| < \sqrt{2} - 1$.

§15. The formulae of Murphy and Rodrigues

From the expansion (13.2) it follows immediately that

$$\begin{aligned}\sum_{n=0}^{\infty} h^n \frac{\mathrm{d}^r}{\mathrm{d}\mu^r} P_n(\mu) &= \frac{\mathrm{d}^r}{\mathrm{d}\mu^r}(1-2\mu h + h^2)^{-\frac{1}{2}} \\ &= 2^r h^r \frac{\Gamma(r+\frac{1}{2})}{\Gamma(\frac{1}{2})}(1-2\mu h+h^2)^{-r-\frac{1}{2}}.\end{aligned}$$

Substituting the value $\mu = 1$ we see that

$$\begin{aligned}\sum_{n=0}^{\infty} P_n^{(r)}(1)h^n &= 2^r \frac{\Gamma(r+\frac{1}{2})}{\Gamma(\frac{1}{2})} h^r (1-h)^{-(2r+1)} \\ &= 2^r \frac{\Gamma(r+\frac{1}{2})}{\Gamma(\frac{1}{2})} h^r \sum_{s=0}^{\infty} \frac{\Gamma(1+2r+s)}{\Gamma(1+2r)s!} h^s,\end{aligned}$$

where $P_n^{(r)}(\mu)$ denotes $\mathrm{d}^r P_n(\mu)/\mathrm{d}\mu^r$.

Equating coefficients of h^n we see that $P_n^{(r)}(1) = 0$ if $r > n$, as is obvious from the fact that $P_n(\mu)$ is a polynomial of degree n in μ, and that

$$P_n^{(r)}(1) = 2^r \frac{\Gamma(r+\frac{1}{2})}{\Gamma(\frac{1}{2})} \frac{\Gamma(1+n+r)}{\Gamma(1+2r)(n-r)!},$$

From the duplication formula for the gamma function

$$\frac{2^r \Gamma(r+\frac{1}{2})}{\Gamma(\frac{1}{2})\Gamma(1+2r)} = \frac{1}{r!2^r} = \frac{1}{(1)_r 2^r},$$

and from Problem 1.27(iii)

$$\frac{\Gamma(1+n+r)}{(n-r)!}=(-1)^r(n+1)_r(-n)_r,$$

so that

$$P_n^{(r)}(1)=(-1)^r\frac{(n+1)_r(-n)_r}{(1)_r 2r}. \tag{15.1}$$

Now, by Taylor's theorem

$$P_n(\mu)=\sum_{r=0}^{\infty}\frac{(\mu-1)^r}{r!}P_n^{(r)}(1).$$

Substituting the expression (15.1) in this expansion we obtain the relation

$$P_n(\mu)=\sum_{r=0}^{\infty}\frac{(-n)_r(n+1)_r}{(1)_r r!}\left(\frac{1-\mu}{2}\right)^r$$

which gives **Murphy's formula,**

$$P_n(\mu)={}_2F_1\left(-n,\, n+1;\, 1;\, \frac{1-\mu}{2}\right), \tag{15.2}$$

for the Legendre polynomial $P_n(\mu)$.

If now we put $\alpha=1$ in Problem 1.5, we see that equation (15.2) is equivalent to

$$P_n(\mu)=\frac{1}{2^n n!}\frac{\mathrm{d}^n}{\mathrm{d}\mu^n}(\mu^2-1)^n, \tag{15.3}$$

which is **Rodrigues' formula** for the Legendre polynomial.

Rodrigues' formula is of great use in the evaluation of definite integrals involving Legendre polynomials. Consider, for instance, the integral

$$I=\int_{-1}^{1} f(x)P_n(x)\,\mathrm{d}x. \tag{15.4}$$

By Rodrigues' formula we may write this integral as

$$\frac{1}{2^n!}\int_{-1}^{1} f(x)\frac{\mathrm{d}^n}{\mathrm{d}x^n}(x^2-1)^n\,\mathrm{d}x,$$

and an integration by parts gives

$$\frac{1}{2^n!}\left[\frac{d^{n-1}}{dx^{n-1}}(x^2-1)^n\right]_{-1}^{1}-\frac{1}{2^n n!}\int_{-1}^{1} f'(x)\frac{d^{n-1}}{dx^{n-1}}\{(x^2-1)^n\}\,dx.$$

The square bracket vanishes at both limits so that we have

$$I=-\frac{1}{2^n n!}\int_{-1}^{1} f'(x)\frac{d^{n-1}}{dx^{n-1}}\{(x^2-1)^n\}\,dx.$$

Continuing this process we find that

$$I=\frac{(-1)^n}{2^n n!}\int_{-1}^{1}(x^2-1)^n f^{(n)}(x)\,dx. \tag{15.5}$$

For example if $f(x)=P_m(x)$, $m<n$, $f^{(n)}(x)=0$ and so $I=0$. In other words

$$\int_{-1}^{1} P_m(x)P_n(x)\,dx=0, \qquad (m\neq n). \tag{15.6}$$

If $f(x)=P_n(x)$ then

$$\begin{aligned} f^{(n)}(x)&=\frac{1}{2^n n!}\frac{d^{2n}}{dx^{2n}}(x^2-1)^n \\ &=\frac{(2n)!}{2^n n!}. \end{aligned}$$

Hence

$$\begin{aligned} \int_{-1}^{1}\{P_n(x)\}^2\,dx&=\frac{(2n)!}{2^{2n}(n!)^2}\int_{-1}^{1}(1-x^2)^n\,dx \\ &=\frac{(2n)!}{2^{2n}(n!)^2}\frac{\Gamma(\frac{1}{2})\Gamma(n+1)}{\Gamma(n+\frac{3}{2})}. \end{aligned}$$

Making use of the duplication formula for the gamma function we can reduce this to the form

$$\int_{-1}^{1}\{P_n(x)\}^2\,dx=\frac{2}{2n+1}. \tag{15.7}$$

A convenient way of combining the results (15.6) and (15.7) is to write

$$\int_{-1}^{1} P_m(x)P_n(x)\,dx=\frac{2}{2n+1}\,\delta_{m,n}, \tag{15.8}$$

where $\delta_{m,n}$ is the Kronecker delta which takes the value 0 if $m \neq n$ and the value 1 if $m = n$.

Similarly if $f(x) = x^m$, where m is a positive integer, then

$$f^{(n)}(x) = \begin{cases} \dfrac{\Gamma(m+1)}{\Gamma(m-n+1)} x^{m-n}, & \text{if} \quad m \geqslant n, \\ 0, & \text{if} \quad m < n, \end{cases}$$

and hence, if $m > n$,

$$\int_{-1}^{1} x^m P_n(x)\,dx = \frac{\Gamma(m+1)}{2^n \Gamma(m-n+1) n!} \int_{-1}^{1} x^{m-n}(1-x^2)^n\,dx.$$

If $m - n$ is an odd integer the integral on the right is zero while if $m - n$ is an even integer it has the value

$$2\int_{0}^{1} x^{m-n}(1-x^2)^n\,dx = \frac{\Gamma(\tfrac{1}{2}m - \tfrac{1}{2}n + \tfrac{1}{2})\Gamma(n+1)}{\Gamma(\tfrac{1}{2}m + \tfrac{1}{2}n + \tfrac{3}{2})},$$

so that, if m is an integer,

$$\int_{-1}^{1} x^m P_n(x)\,dx$$

$$= \begin{cases} 0, & \text{if } m < n, \\ \dfrac{m!\,\Gamma(\tfrac{1}{2}m - \tfrac{1}{2}n + \tfrac{1}{2})}{2^n (m-n)!\,\Gamma(\tfrac{1}{2}m + \tfrac{1}{2}n + \tfrac{3}{2})}, & \text{if } m - n \geqslant 0 \text{ is even,} \\ 0, & \text{if } m - n > 0 \text{ is odd.} \end{cases} \qquad (15.9)$$

If $m = n$ the result is

$$\int_{-1}^{1} x^n P_n(x)\,dx = \frac{1}{2^n}\int_{-1}^{1} (1-x^2)^n\,dx$$

$$= \frac{1}{2^n}\frac{\Gamma(\tfrac{1}{2})\Gamma(n+1)}{\Gamma(n+\tfrac{3}{2})}$$

which, on account of the duplication formula, is equivalent to

$$\int_{-1}^{1} x^n P_n(x)\,dx = \frac{2^{n+1}(n!)^2}{(2n+1)!}. \qquad (15.10)$$

§16. Series of Legendre polynomials

In certain problems of potential theory it is desirable to be able to express a given function in the form of a series of Legendre polynomials. We can readily show that this is possible in the case in which the given function is a simple polynomial. For example, from the equations (13.4a) we have

$$\begin{aligned}
1 &= P_0(\mu), \\
\mu &= P_1(\mu), \\
\mu^2 &= \tfrac{1}{3} + \tfrac{2}{3}P_2(\mu) = \tfrac{1}{3}P_0(\mu) + \tfrac{2}{3}P_2(\mu), \\
\mu^3 &= \tfrac{3}{5}\mu + \tfrac{2}{5}P_3(\mu) = \tfrac{3}{5}P_1(\mu) + \tfrac{2}{5}P_3(\mu),
\end{aligned}$$

so that any cubic $c_0\mu^3 + c_1\mu^2 + c_2\mu + c_3$ can be written as the series

$$\frac{2c_0}{5}P_3(\mu) + \frac{2c_1}{3}P_2(\mu) + \left(\frac{3c_0}{5} + c_2\right)P_1(\mu) + \left(\frac{2c_1}{3} + c_3\right)P_0(\mu).$$

It is obvious that we could proceed in this way for a polynomial of any given degree n, though if n were large the arithmetic involved might become cumbersome. Since $P_n(\mu)$ is a polynomial of degree n in μ, it emerges as a result of an extension of the above argument that *any* polynomial of degree n in μ can be expressed as a series of the type

$$\sum_{r=0}^{n} c_r P_r(\mu), \qquad -1 \leqslant \mu \leqslant 1. \tag{16.1}$$

The problem which now arises is that of expressing an arbitrary function $f(\mu)$ defined on $[-1, 1]$ as an infinite series of Legendre polynomials. From equation (15.8) we see that the set of polynomials $\{\Psi_n\}_{n=1}^{\infty}$ defined by

$$\Psi_n(\mu) = (n + \tfrac{1}{2})^{\frac{1}{2}} P_n(\mu) \tag{16.2}$$

is orthonormal on $[-1, 1]$. From equation (4.14) we then deduce that

$$f(\mu) = \sum_{n=0}^{\infty} c_n P_n(\mu), \qquad |\mu| < 1, \tag{16.3}$$

where

$$c_n = (n + \tfrac{1}{2}) \int_{-1}^{1} f(\mu) P_n(\mu)\, d\mu. \tag{16.4}$$

The series on the right side of equation (16.3) is called the **Fourier–Legendre series** of the function f; the constants c_n, defined by equation (16.4) are called the **Fourier–Legendre coefficients of f.**

We shall not discuss here the conditions which must be satisfied by the function f if its Fourier–Legendre is to be uniformly convergent; for such a discussion the reader is referred to Chapter VII of Hobson (1965).

§17. Legendre's differential equation

If we write

$$v=(\mu^2-1)^n,$$

then it is readily shown that

$$(1-\mu^2)\frac{dv}{d\mu}+2\mu nv=0$$

and if we differentiate this equation $n+1$ times using Leibnitz's theorem we find that

$$(1-\mu^2)\frac{d^{n+2}v}{d\mu^{n+2}}-2\mu\frac{d^{n+1}v}{d\mu^{n+1}}+n(n+1)\frac{d^n v}{d\mu^n}=0,$$

which when written in the form

$$\left\{(1-\mu^2)\frac{d^2}{d\mu^2}-2\mu\frac{d}{d\mu}+n(n+1)\right\}\left(\frac{d^n v}{d\mu^n}\right)=0$$

shows that $d^n(\mu^2-1)^n/d\mu^n$ is a solution of the differential equation

$$(1-\mu^2)\frac{d^2y}{d\mu^2}-2\mu\frac{dy}{d\mu}+n(n+1)y=0, \tag{17.1}$$

so that we conclude from Rodrigues' formula (15.3) that when n is an integer $P_n(\mu)$ is one solution of the equation (17.1). This equation, which we shall now consider in a little more detail, is called **Legendre's differential equation.** We saw in Problem 1.1 how such an equation arises in the solution of Laplace's equation when solutions of the type $R(r)\Theta(\cos\theta)$, i.e. $m=0$, are sought.

It is obvious by inspection that the point $\mu=0$ is an ordinary point of the equation (17.1). Writing the equation in the form

$$(\mu-1)^2\frac{d^2y}{d\mu^2}+(\mu-1)\frac{2\mu}{\mu+1}\frac{dy}{d\mu}-\frac{n(n+1)}{\mu+1}(\mu-1)y=0$$

and observing that in the notation of equation (3.1) with $a=1$,

$$p(\mu)=\frac{1+(\mu-1)}{1+\frac{1}{2}(\mu-1)}=1+\tfrac{1}{2}(\mu-1)-\tfrac{1}{4}(\mu-1)^2\ldots,$$

$$q(\mu)=-\tfrac{1}{2}n(n+1)\frac{\mu-1}{1+\frac{1}{2}(\mu-1)}$$
$$=-\tfrac{1}{2}n(n+1)\{(\mu-1)-\tfrac{1}{2}(\mu-1)^2+\ldots\},$$

we see that $\mu=1$ is a regular singular point with indicial equation $\rho^2=0$.

Furthermore in the notation of equation (4.2)

$$\alpha(\mu)\equiv-\frac{2\mu}{1-\mu^2},\qquad \beta(\mu)=\frac{n(n+1)}{1-\mu^2},$$

so that as $\mu\to\infty$

$$\alpha(\mu)\sim\frac{2}{\mu},\qquad \beta(\mu)\sim-\frac{n(n+1)}{\mu^2},$$

showing that the point $\mu=\infty$ is a regular singular point with indicial equation $\{\rho-(n+1)\}\,(\rho+n)=0$. We thus see that the equation is defined by the scheme

$$y=P\begin{Bmatrix}-1 & \infty & 1 & \\ 0 & n+1 & 0 & \mu \\ 0 & -n & 0 & \end{Bmatrix}. \tag{17.2}$$

If, however, we put

$$x=\tfrac{1}{2}(1-\mu)$$

in equation (17.1), we find that it reduces to the form

$$x(1-x)\frac{d^2y}{dx^2}+(1-2x)\frac{dy}{dx}+n(n+1)y=0, \tag{17.3}$$

which is equation (8.1) with $\alpha=n+1$, $\beta=-n$ and $\gamma=1$, so that the scheme (17.2) is equivalent to the scheme

$$y=P\begin{Bmatrix}0 & \infty & 1 & \\ 0 & n+1 & 0 & \tfrac{1}{2}-\tfrac{1}{2}\mu \\ 0 & -n & 0 & \end{Bmatrix}. \tag{17.4}$$

It should be noticed that the values along the top row are those assumed by $\frac{1}{2}-\frac{1}{2}\mu$, not by μ.

We consider first the solution corresponding to the singular point at infinity. We write

$$y_1(\mu)=\mu^n \sum_{\nu=0}^{\infty} c_\nu \mu^{-\nu}$$

which on substitution into (17.1) leads to the recurrence relation

$$(n-\nu+2)(n-\nu+1)c_{\nu-2}=-\nu(2n+1-\nu)c_\nu.$$

On taking $c_0=1$ we obtain the solution

$$y_1(\mu)=\mu^n-\frac{n(n-1)}{2.(2n-1)}\mu^{n+2}+\frac{n(n-1)(n-2)(n-3)}{2.4.(2n-1)(2n-3)}\mu^{n-4}+\dots$$

$$=\mu^n\left\{1+\frac{(-\tfrac{1}{2}n)(\tfrac{1}{2}-\tfrac{1}{2}n)}{1.(\tfrac{1}{2}-n)}\frac{1}{\mu^2}\right.$$

$$\left.+\frac{(-\tfrac{1}{2}n)(-\tfrac{1}{2}n+1)(\tfrac{1}{2}-\tfrac{1}{2}n)(\tfrac{1}{2}-\tfrac{1}{2}n+1)}{1.2(\tfrac{1}{2}-n)(\tfrac{1}{2}-n+1)}\frac{1}{\mu^4}+\dots\right\},$$

which may be written in the form

$$y_1(\mu)=\mu^n\,{}_2F_1\left(-\tfrac{1}{2}n,\tfrac{1}{2}-\tfrac{1}{2}n;\tfrac{1}{2}-n;\frac{1}{\mu^2}\right). \tag{17.5}$$

Also, if we write for the second solution

$$y_2(\mu)=\mu^{-n-1}\sum_{\nu=0}^{\infty} d_\nu \mu^{-\nu}$$

we find that

$$y_2(\mu)=\mu^{-n-1}+\frac{(n+1)(n+2)}{2.(2n+3)}\mu^{-n-3}$$

$$+\frac{(n+1)(n+2)(n+3)(n+4)}{2.4.(2n+3)(2n+5)}\mu^{-n-5}+\dots$$

$$=\frac{1}{\mu^{n+1}}\,{}_2F_1\left(\tfrac{1}{2}n+\tfrac{1}{2},\tfrac{1}{2}n+1;n+\tfrac{3}{2};\frac{1}{\mu^2}\right), \tag{17.6}$$

provided n is any number other than a negative integer or half a negative integer.

These solutions are valid for *all* values of n for which the ${}_2F_1$ series have a meaning, not only for integral values ($|\mu|>1$). In some problems we know that the solution should be a polynomial in μ; in that case we must take the solution to the form $y=AP_n(\mu)$.

The variation of $Q_n(\mu)$ with μ may be computed easily from

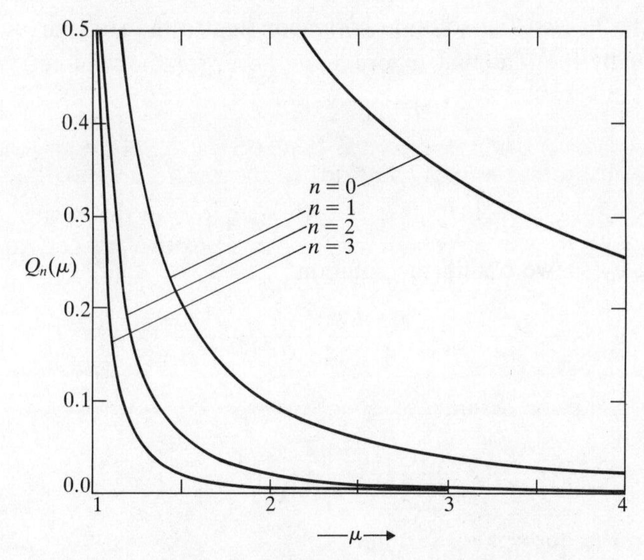

Fig. 7 Variation of $Q_n(\mu)$ with μ

equation (17.7) when $\mu > 1$. Tables calculated in this way are contained in the volume quoted at the end of §13. Figure 7, which was prepared from these tables, shows the variation of $Q_n(\mu)$ with μ for a few values of n.

Since Legendre's equation has a regular singular point at $\mu = 1$ we may on the basis of equation (3.8) take the second solution of Legendre's equation to be proportional to

$$P_n(\mu) \log(\mu - 1) + \Sigma c_r(\mu - 1)^r.$$

The coefficients c_r can now be obtained by substituting this expression into the differential equation (17.1) and equating to zero coefficients of successive powers of $(\mu - 1)$. Instead of proceeding in this way we shall derive this second solution by means of a method due to F. E. Neumann.

§18. Neumann's formula for the Legendre functions

Let us now consider the integral

$$\int_{-1}^{1} \frac{P_n(\xi)\, d\xi}{\mu - \xi},$$

where $|\mu|>1$, and n is a positive integer. Expanding the denominator by the binomial theorem, we have, for the value of the integral, the series

$$\sum_{s=0}^{\infty} \frac{1}{\mu^{s+1}} \int_{-1}^{1} \xi^s P_n(\xi)\, d\xi.$$

From equation (15.9), it follows that the integrals in this series are zero if $s \leqslant n$, or if $s=n+2r$ where r is a positive integer, so that the above series is equivalent to

$$\sum_{r=0}^{\infty} \frac{1}{\mu^{n+1+2r}} \int_{-1}^{1} \xi^{n+2r} P_n(\xi)\, d\xi,$$

which, by the same formula, is equal to

$$\frac{1}{\mu^{n+1}} \sum_{r=0}^{\infty} \frac{(n+2r)!\Gamma(r+\frac{1}{2})}{2^n(2r)!\Gamma(n+r+\frac{3}{2})} \left(\frac{1}{\mu^2}\right)^r.$$

Now from the duplication formula

$$\frac{(n+2r)!\Gamma(\frac{1}{2}+r)}{2^n(2r)!} = \frac{\Gamma(\frac{1}{2}n+\frac{1}{2}+r)\Gamma(\frac{1}{2}n+1+r)}{r!},$$

so that the series reduces to

$$\frac{1}{\mu^{n+1}} \sum_{r=0}^{\infty} \frac{\Gamma(\frac{1}{2}n+\frac{1}{2}+r)\Gamma(\frac{1}{2}n+1+r)}{\Gamma(n+\frac{3}{2}+r)r!} \left(\frac{1}{\mu^2}\right)^r,$$

which is equal to

$$\frac{\Gamma(\frac{1}{2}n+\frac{1}{2})\Gamma(\frac{1}{2}n+1)}{\Gamma(n+\frac{3}{2})\mu^{n+1}}\, {}_2F_1\left(\tfrac{1}{2}n+\tfrac{1}{2}, \tfrac{1}{2}n+1; n+\tfrac{3}{2}; \frac{1}{\mu^2}\right).$$

Noticing that

$$\Gamma(\tfrac{1}{2}n+\tfrac{1}{2})\Gamma(\tfrac{1}{2}n+1)/\Gamma(n+\tfrac{3}{2}) = \Gamma(\tfrac{1}{2})\Gamma(n+1)/2^n\Gamma(n+\tfrac{3}{2}),$$

and comparing the result with equation (17.7) we see that this series is merely $2Q_n(\mu)$. Hence we have shown that if $|\mu|>1$,

$$Q_n(\mu) = \frac{1}{2}\int_{-1}^{1} \frac{P_n(\xi)}{\mu-\xi}\, d\xi, \tag{18.1}$$

a result which is known as **Neumann's formula.** Equation (18.1) holds not only for real values of μ greater than 1, but for all values of μ which are not real. For this reason equation (18.1) may be regarded as defining the second solution, $Q_n(\mu)$, of Legendre's equation.

Certain related formulae, due to MacRobert, follow readily from this result. If $|\mu|>1$ and m is a positive integer, then

$$\mu^m Q_n(\mu)-\frac{1}{2}\int_{-1}^{1}\frac{\xi^m P_n(\xi)}{\mu-\xi}\,d\xi=\frac{1}{2}\int_{-1}^{1}\frac{\mu^m-\xi^m}{\mu-\xi}P_n(\xi)\,d\xi,$$

and the integral on the right is equivalent to the finite sum

$$\tfrac{1}{2}\sum_{r=0}^{m-1}\mu^{m-1-r}\int_{-1}^{1}\xi^r P_n(\xi)\,d\xi. \tag{18.2}$$

If $m\leqslant n$ it follows from equation (15.9) that each term of this series vanishes so that we have

$$\mu^m Q_n(\mu)=\frac{1}{2}\int_{-1}^{1}\frac{\xi^m P_n(\xi)}{\mu-\xi}\,d\xi \tag{18.3}$$

provided m, n are integers and $m\leqslant n$.

On the other hand, if $m=n+1$ the series (18.2) reduces to the single term

$$\frac{1}{2}\int_{-1}^{1}\xi^n P_n(\xi)\,d\xi=\frac{2^n(n!)^2}{(2n+1)!},$$

so that

$$\mu^{n+1}Q_n(\mu)=\frac{1}{2}\int_{-1}^{1}\frac{\xi^{n+1}P_n(\xi)}{\mu-\xi}\,d\xi+\frac{2^n(n!)^2}{(2n+1)!}. \tag{18.4}$$

Now if m is an integer $P_m(\mu)$ is a polynomial of degree m in μ, so that it follows from (18.3) by finite summation that, if $m\leqslant n$,

$$P_m(\mu)Q_n(\mu)=\frac{1}{2}\int_{-1}^{1}\frac{P_m(\xi)P_n(\xi)}{\mu-\xi}\,d\xi. \tag{18.5}$$

Similarly from equation (18.4) and the definition of $P_{n+1}(\mu)$, we have the formula

$$P_{n+1}(\mu)Q_n(\mu)=\frac{1}{2}\int_{-1}^{1}\frac{P_{n+1}(\xi)P_n(\xi)}{\mu-\xi}\,d\xi+\frac{1}{n+1}. \tag{18.6}$$

If we replace n in equation (18.5) by $n+1$ and m by n and subtract from equation (18.6) we obtain the relation

$$P_{n+1}(\mu)Q_n(\mu)-P_n(\mu)Q_{n+1}(\mu)=\frac{1}{n+1}. \tag{18.7}$$

Other formulae of a similar nature are contained in Problem 3.25 below.

We shall now make use of Neumann's formula to derive the form of the second solution of Legendre's equation in the neighbourhood of the points $\mu=\pm 1$. From (18.1) we have the result that if $\mu>1$,

$$Q_n(\mu)=\tfrac{1}{2}P_n(\mu)\log\frac{\mu+1}{\mu-1}-W_{n-1}(\mu), \tag{18.8}$$

where $W_{n-1}(\mu)$ denotes the integral

$$\frac{1}{2}\int_{-1}^{1}\frac{P_n(\mu)-P_n(\xi)}{\mu-\xi}\,d\xi.$$

Now when n is an integer, $P_n(\mu)$ is a polynomial of degree n in μ so that $\{P_n(\mu)-P_n(\xi)\}/(\mu-\xi)$ is a polynomial of degree $n-1$ in μ. Hence $W_{n-1}(\mu)$ is a polynomial of degree $n-1$ in μ.

If we substitute from (18.8) into Legendre's equation (17.1), we find that W_{n-1} satisfies the differential equation

$$(1-\mu^2)W''_{n-1}-2\mu W'_{n-1}+n(n+1)W_{n-1}=2P'_n(\mu). \tag{18.9}$$

Now since $W_{n-1}(\mu)$ is a polynomial of degree $n-1$ in μ we may write (cf. §16 above)

$$W_{n-1}(\mu)=\sum_{r=0}^{n-1}c_rP_r(\mu).$$

Using the fact that

$$P'_n(\mu)=\sum_{r=0}^{p}(2n-4r-1)P_{n-2r-1},$$

where $p=\tfrac{1}{2}(n-1)$ or $\tfrac{1}{2}n-1$ according as n is odd or even (which follows from equation (14.6)), and the result

$$(1-\mu^2)P''_r(\mu)-2\mu P'_r(\mu)+n(n+1)P_r(\mu)=(n-r)(n+r+1)P_r(\mu),$$

we find, from equation (18.9), that $c_{n-2s}=0$ $(s=0,\ldots,p)$, and that

$$c_{n-2s-1}=\frac{2n-4s-1}{(2s+1)(n-s)}.$$

Substituting these values in (18.8) we obtain the formula

$$Q_n(\mu)=\tfrac{1}{2}P_n(\mu)\log\frac{\mu+1}{\mu-1}-\sum_{s=0}^{p}\frac{2n-4s-1}{(2s+1)(n-s)}P_{n-2s-1}(\mu). \tag{18.10}$$

Another expression for $Q_n(\mu)$ may be derived from the fact that both $P_n(\mu)$ and $Q_n(\mu)$ are solutions of Legendre's equation (17.1), so that

$$Q_n(\mu)\frac{\mathrm{d}}{\mathrm{d}\mu}\{(1-\mu^2)P_n'(\mu)\}-P_n(\mu)\frac{\mathrm{d}}{\mathrm{d}\mu}\{(1-\mu^2)Q_n'(\mu)\}=0,$$

which is equivalent to

$$\frac{\mathrm{d}}{\mathrm{d}\mu}[(1-\mu^2)\{Q_n(\mu)P_n'(\mu)-P_n(\mu)Q_n'(\mu)\}]=0,$$

showing that

$$(1-\mu^2)\{Q_n(\mu)P_n'(\mu)-P_n(\mu)Q_n'(\mu)\}=C, \qquad (18.11)$$

where C is a constant. Now from equations (17.7) and (17.8), we can readily show that for large values of μ

$$P_n'(\mu)Q_n(\mu)\sim\frac{n}{2n+1}\frac{1}{\mu^2}, \qquad Q_n'(\mu)P_n(\mu)\sim-\frac{n+1}{2n+1}\frac{1}{\mu^2},$$

so that as $\mu\to\infty$ the left-hand side of (18.11) tends to -1, showing that $C=-1$. Writing (18.11) in the form

$$\frac{\mathrm{d}}{\mathrm{d}\mu}\left\{\frac{Q_n(u)}{P_n(\mu)}\right\}=\frac{-1}{(\mu^2-1)\{P_n(\mu)\}^2}$$

and noting from (17.7) that $Q_n(\mu)\to 0$ as $\mu\to\infty$, we have

$$Q_n(\mu)=P_n(\mu)\int_\mu^\infty\frac{\mathrm{d}\xi}{(\xi^2-1)\{P_n(\xi)\}^2}. \qquad (18.12)$$

§19. Recurrence relations for the function $Q_n(\mu)$

Recurrence relations for the Legendre function of the second kind can be derived from Neumann's formula (18.1) and the corresponding recurrence relations for the Legendre polynomials $P_n(\mu)$. From the recurrence relation (14.2) and Neumann's formula we have

$$(n+1)Q_{n+1}(\mu)+nQ_{n-1}(\mu)=(n+\tfrac{1}{2})\int_{-1}^{1}\frac{\xi P_n(\xi)}{\mu-\xi}\,\mathrm{d}\xi.$$

Now

$$\frac{1}{2}\int_{-1}^{1} \frac{\xi P_n(\xi)}{\mu - \mu}\, d\xi = \mu Q_n(\mu) - \tfrac{1}{2}\int_{-1}^{1} P_n(\xi)\, d\xi.$$

If we write the second term on the right as $\int_{-1}^{1} P_0(\xi)P_n(\xi)\, d\xi$, we see from (15.6) that it vanishes if $n \neq 0$. Hence we have

$$(n+1)Q_{n+1}(\mu) - (2n+1)\mu Q_n(\mu) + nQ_{n-1}(\mu) = 0,$$

showing that the functions $Q_n(\mu)$ for three consecutive values of n satisfy a relation of the same form as that for the functions $P_n(\mu)$ (equation (14.2) above).

From Neumann's formula (18.1) we have

$$Q_n'(\mu) = -\frac{1}{2}\int_{-1}^{1} \frac{P_n(\xi)}{(\mu-\xi)^2}\, d\xi$$

and if we integrate by parts on the right-hand side, we find that

$$Q_n'(\mu) = \frac{1}{2}\left\{\frac{1}{1-\mu} + \frac{(-1)^n}{1+\mu}\right\} + \frac{1}{2}\int_{-1}^{1} \frac{P_n'(\xi)}{\mu - \xi}\, d\xi.$$

Hence

$$\begin{aligned} Q_{n+1}'(\mu) - Q_{n-1}'(\mu) &= \frac{1}{2}\int_{-1}^{1} \frac{P_{n+1}'(\xi) - P_{n-1}'(\xi)}{\mu - \xi}\, d\xi \\ &= \tfrac{1}{2}(2n+1)\int_{-1}^{1} \frac{P_n(\xi)}{\mu - \xi}\, d\xi, \end{aligned}$$

by virtue of equation (14.6). The integral on the right is $2Q_n(\mu)$ by Neumann's formula so that, finally,

$$Q_{n+1}'(\mu) - Q_{n-1}'(\mu) = (2n+1)Q_n(\mu).$$

§20. The use of Legendre functions in potential theory

In potential theory we have frequently to determine solutions of Laplace's equation $\nabla^2\psi = 0$ which satisfy certain prescribed boundary conditions. If we have a problem in which the natural boundaries are spheres with centre at the origin of coordinates, it is natural to employ polar coordinates r, θ, ϕ. In cases in which there is symmetry about the polar axis, ψ will not depend on ϕ

so we may write $\psi=\psi(r,\theta)$. It then follows from Problem 1.1 that

$$(A_n r^n + B_n r^{-n-1})v_n$$

will be a solution of Laplace's equation provided that v_n is a solution of Legendre's equation (17.1). Taking v_n to be $P_n(\cos\theta)+C_nQ_n(\cos\theta)$, we see that we may write

$$\psi(r,\theta)=\sum_{n=0}^{\infty}(A_n r^n+B_n r^{-n-1})P_n(\cos\theta)$$
$$+\sum_{n=0}^{\infty}(C_n r^n+D_n r^{-n-1})Q_n(\cos\theta), \qquad (20.1)$$

where the quantities A_n, B_n, C_n, D_n $(n=0,1,2,\ldots)$ are all constants.

Now it is obvious from equation (18.8) that $Q_n(\cos\theta)$ is infinite when $\theta=0^0$, and we know on physical grounds that in the case of spherical boundaries ψ remains finite along the axis $\theta=0$. Hence we must take $C_n=D_n=0$ for all values of n and obtain the potential function

$$\psi=\sum_{n=0}^{\infty}(A_n r^n+B_n r^{-n-1})P_n(\cos\theta), \qquad (20.2)$$

which is valid as long as r is neither zero nor infinite, i.e. if $a\leqslant r\leqslant b$ where a and b are finite and non-zero. If the region under discussion is the interior of a sphere, i.e. if $0\leqslant r\leqslant a$, then to avoid ψ becoming infinite we must take B_n to be zero to give

$$\psi=\sum_{n=0}^{\infty}A_n r^n P_n(\cos\theta). \qquad (20.3)$$

On the other hand, if the region being considered lies entirely outside this sphere, we must take

$$\psi=\sum_{n=0}^{\infty}B_n r^{-n-1}P_n(\cos\theta). \qquad (20.4)$$

Examples of the use of the solutions (20.2, 3, 4) in potential theory are given in Coulson and Boyd (1979), pp. 159–70; a further example is given below.

The Legendre functions of the second kind, $Q_n(\cos\theta)$, which are absent from problems involving spherical boundaries, enter into the expressions for potential functions appropriate to the space between two coaxial cones. If $0<\alpha<\theta<\beta<\pi$, we must take a solution of the form (20.1). Suppose, for example, that $\psi=0$ on $\theta=\alpha$, and $\psi=\sum \alpha_n r^n$ on $\theta=\beta$, then we must have

$$A_n P_n(\cos\alpha)+C_n Q_n(\cos\alpha)=0$$

and

$$A_n P_n(\cos\beta)+C_n Q_n(\cos\beta)=\alpha_n, \qquad B_n=D_n=0,$$

the latter results following from the fact that if $\alpha\neq\beta$,

$$Q_n(\cos\alpha)P_n(\cos\beta)-P_n(\cos\alpha)Q_n(\cos\beta)$$

does not vanish. Solving these equations for A_n and C_n and inserting the solutions in equation (20.1), we find that in the space between the two cones

$$\psi=\sum_{n=0}^{\infty}\alpha_n r^n\left\{\frac{Q_n(\cos\alpha)P_n(\cos\theta)-P_n(\cos\alpha)Q_n(\cos\theta)}{Q_n(\cos\alpha)P_n(\cos\beta)-P_n(\cos\alpha)Q_n(\cos\beta)}\right\}. \tag{20.5}$$

To illustrate the use of the solution (20.1) and of some of the properties of Legendre functions, we shall now consider the problem in which an insulated conducting sphere of radius a is placed with its centre at the origin of coordinates in an electric field whose potential is known to be

$$\sum_{n=1}^{\infty}\alpha_n r^n P_n(\cos\theta) \tag{20.6}$$

and we wish to determine the force on the sphere. The conditions to be satisfied by the potential functions ψ are (i) that ψ is a solution of Laplace's equation; (ii) that ψ has the form (20.6) for large values of r; (iii) $\psi=0$ on $r=a$.

The conditions (i) and (ii) are satisfied if we take

$$\psi=\sum_{n=1}^{\infty}\left(\alpha_n r^n+\frac{B_n}{r^{n+1}}\right)P_n(\cos\theta)$$

and (iii) is satisfied if we write $B_n=-\alpha_n a^{2n+1}$. We therefore have

$$\psi=\sum_{n=1}^{\infty}\alpha_n\left(r^n-\frac{a^{2n+1}}{r^{n+1}}\right)P_n(\cos\theta).$$

The surface density of charge on the conductor is

$$\sigma=-\frac{1}{4\pi}\left(\frac{\partial\psi}{\partial r}\right)_{r=a}=-\frac{1}{4\pi}\sum_{n=1}^{\infty}(2n+1)a^{n-1}\alpha_n P_n(\cos\theta),$$

and since the force per unit area on the conductor is $2\pi\sigma^2$, the resultant force on the sphere is in the $\theta = 0$ direction and is of magnitude

$$F = \int_0^\pi 2\pi\sigma^2 2\pi a^2 \sin\theta \cos\theta \, d\theta$$
$$= \tfrac{1}{4}a^2 \sum_{n=1}^{\infty} \sum_{m=1}^{\infty} \alpha_m \alpha_n a^{m+n-2} I_{mn}, \tag{20.7}$$

where I_{mn} denotes the integral

$$(2n+1)(2m+1)\int_0^\pi \cos\theta \sin\theta \, P_m(\cos\theta) P_n(\cos\theta) \, d\theta.$$

Changing the variable of integration to $\mu = \cos\theta$ and using the recurrence relation (14.2), we find that

$$I_{mn} = (2m+1)\left\{(n+1)\int_{-1}^{1} P_m(\mu)P_{n+1}(\mu)\, d\mu + n\int_{-1}^{1} P_m(\mu)P_{n-1}(\mu)\, d\mu\right\},$$

and by the orthogonality property (15.8) this reduces to the form

$$I_{mn} = 2(n+1)\delta_{m,n+1} + 2n\delta_{m,n-1}. \tag{20.8}$$

Substituting from (20.8) into (20.7) we find that the total force on the sphere is

$$F = \sum_{n=1}^{\infty} (n+1)\alpha_n \alpha_{n+1} a^{2n+1}.$$

§21. Legendre's associated functions

We saw in Problem 1.1 that the solution of Laplace's equation in spherical polar coordinates reduces to the solution of the ordinary differential equation

$$(1-\mu^2)\frac{d^2\Theta}{d\mu^2} - 2\mu\frac{d\Theta}{d\mu} + \left\{n(n+1) - \frac{m^2}{1-\mu^2}\right\}\Theta = 0, \tag{21.2}$$

which reduces to Legendre's equation when $m = 0$. This equation is known as **Legendre's associated equation.** To solve this equation we may write

$$\Theta = (\mu^2 - 1)^{-\frac{1}{2}m} y. \tag{21.2}$$

Substituting this expression in the differential equation we find that y satisfies the equation

$$(1-\mu^2)\frac{d^2y}{d\mu^2}-2(1-m)\mu\frac{dy}{d\mu}+(n+m)(n-m+1)y=0,$$

and differentiating this equation m times with respect to μ by Leibnitz's theorem we find that

$$\left\{(1-\mu^2)\frac{d^2}{d\mu^2}-2\mu\frac{d}{d\mu}+n(n+1)\right\}\frac{d^my}{d\mu^m}=0, \qquad (21.3)$$

showing that if $d^my/d\mu^m$ is a solution of Legendre's equation (17.1) the function Θ, defined by equation (21.2), is a solution of Legendre's associated equation (21.1).

Similarly if we put $\Theta=(\mu^2-1)^{\frac{1}{2}m}y$ in equation (21.1), we find that

$$(1-\mu^2)\frac{d^2y}{d\mu^2}-2(1+m)\mu\frac{dy}{d\mu}+(n-m)(n+m+1)y=0. \qquad (21.4)$$

If now we differentiate equation (17.1) m times with respect to μ we obtain the equation

$$\left\{(1-\mu^2)\frac{d^2}{d\mu^2}-2(1+m)\mu\frac{d}{d\mu}+(n-m)(n+m+1)\right\}\frac{d^my}{d\mu^m}=0,$$

showing that if $y(\mu)$ is a solution of Legendre's equation, then

$$(\mu^2-1)^{\frac{1}{2}m}\frac{d^my}{d\mu^m} \qquad (21.5)$$

is a solution of Legendre's associated equation (21.1).

Taking the two solutions of Legendre's equation to be $P_n(\mu)$ and $Q_n(\mu)$, it follows from (21.5) that the functions

$$P_n^m(\mu)=(\mu^2-1)^{\frac{1}{2}m}\frac{d^mP_n(\mu)}{d\mu^m},$$

$$Q_n^m(\mu)=(\mu^2-1)^{\frac{1}{2}m}\frac{d^mQ_n(\mu)}{d\mu^m} \qquad (21.6)$$

are solutions of Legendre's associated equation. As a consequence of equation (21.3) we see that so also are the functions

$$P_n^{-m}(\mu)=(\mu^2-1)^{-\frac{1}{2}m}\int_1^{\mu}\int_1^{\xi}\cdots\int_1^{\xi}P_n(\xi)(d\xi)^m \qquad (21.7)$$

and

$$Q_n^{-m}(\mu)=(\mu^2-1)^{-\frac{1}{2}m}\int_\infty^\mu\int_\infty^\xi\dots\int_\infty^\xi Q_n(\xi)(\mathrm{d}\xi)^m. \tag{21.8}$$

It is an immediate generalisation of (6.5) that

$$\frac{\mathrm{d}^m}{\mathrm{d}x^m}\,{}_2F_1(\alpha,\beta;\gamma;x)=\frac{(\alpha)_m(\beta)_m}{(\gamma)_m}\,{}_2F_1(\alpha+m,\beta+m;\gamma+m;x),$$

so that using Murphy's form (15.2) for $P_n(\mu)$ we see that

$$\frac{\mathrm{d}^m}{\mathrm{d}\mu^m}P_n(\mu)=\frac{(-n)_m(n+1)_m}{(-2)^m m!}\,{}_2F_1\left(m-n,n+m+1;m+1;\frac{1-\mu}{2}\right).$$

Hence $P_n^m(\mu)$ as defined in (21.6) may be written in the form

$$P_n^m(\mu)=\frac{\Gamma(n+m+1)}{2^m m!\,\Gamma(n-m+1)}(\mu^2-1)^{\frac{1}{2}m}\times{}_2F_1\left(m-n,n+m+1;m+1;\frac{1-\mu}{2}\right). \tag{21.9}$$

Other expressions for the first of the two functions (21.6) can be easily derived. If we make use of the result (7.4) we see that

$$P_n^m(\mu)=\frac{\Gamma(n+m+1)}{2^n m!\,\Gamma(n-m+1)}(\mu+1)^{n-\frac{1}{2}m}(\mu-1)\times{}_2F_1\left(m-n,-n;m+1;\frac{\mu-1}{\mu+1}\right) \tag{21.10}$$

and similarly, if we make use of relation (7.6) we may derive the expression

$$P_n^m(\mu)=\frac{\Gamma(n+m+1)}{m!\,\Gamma(n-m+1)}\left(\frac{\mu-1}{\mu+1}\right)^{\frac{1}{2}m}\times{}_2F_1\left(n+1,-n;m+1;\frac{1-\mu}{2}\right). \tag{21.11}$$

From Rodrigues' formula (15.3) we derive the simple expression

$$P_n^m(\mu)=\frac{1}{2^n n!}(\mu^2-1)^{\frac{1}{2}m}\frac{\mathrm{d}^{m+n}}{\mathrm{d}\mu^{m+n}}(\mu^2-1)^n. \tag{21.12}$$

The simple differentiation

$$\frac{\mathrm{d}^m}{\mathrm{d}\mu^m}\mu^{-n-1-2r}=(-1)^m\frac{\Gamma(m+n+1+2r)}{\Gamma(n+1+2r)}\mu^{-m-n-1-2r}$$

may (because of the duplication formula) be written in the form

$$\Gamma(n+1)\frac{\mathrm{d}^m}{\mathrm{d}\mu^m}\left\{\frac{(\frac{1}{2}n+\frac{1}{2})_r(\frac{1}{2}n+1)_r}{\mu^{n+1+2r}}\right\}$$

$$=(-1)^m\Gamma(m+n+1)\frac{(\frac{1}{2}m+\frac{1}{2}n+\frac{1}{2})_r(\frac{1}{2}m+\frac{1}{2}n+1)_r}{\mu^{m+n+1+2r}},$$

showing that, by term-by-term differentiation of the solution (17.7) of Legendre's equation, we obtain the solution

$$Q_n^m(\mu)=(-1)^m\frac{\Gamma(\frac{1}{2})\Gamma(m+n+1)}{2^{n+1}\Gamma(n+\frac{3}{2})}\mu^{-m-n-1}(\mu^2-1)^{\frac{1}{2}m}$$
$$\times{}_2F_1\left(\tfrac{1}{2}m+\tfrac{1}{2}n+\tfrac{1}{2},\tfrac{1}{2}m+\tfrac{1}{2}n+1;n+\tfrac{3}{2};\frac{1}{\mu^2}\right) \quad (21.13)$$

of Legendre's associated equation.

Solutions of the type (21.7) and (21.8) can be derived in a similar fashion. Using the result

$$\int_1^\mu\int_1^\xi\cdots\int_1^\xi\left(\frac{1-\xi}{2}\right)^r(\mathrm{d}\xi)^m=\frac{1}{m!}(\mu-1)^m\frac{r!}{(m+1)_r}\left(\frac{1-\mu}{2}\right)^r$$

in equation (21.7), with Murphy's expression (15.2) for $P_n(\mu)$, we derive the expression

$$\frac{1}{m!}(\mu-1)^m\,{}_2F_1\left(-n,n+1;m+1;\frac{1-\mu}{2}\right) \quad (21.14)$$

for

$$\int_1^\mu\int_1^\xi\cdots\int_1^\xi P_n(\xi)(\mathrm{d}\xi)^m,$$

and this yields the solution

$$P_n^{-m}(\mu)=\frac{1}{m!}\left(\frac{\mu-1}{\mu+1}\right)^{\frac{1}{2}m}{}_2F_1\left(-n,n+1;m+1;\frac{1-\mu}{2}\right). \quad (21.15)$$

The solution provided by Rodrigues' formula (15.3) can obviously be written as

$$P_n^{-m}(\mu)=\frac{(\mu^2-1)^{-\frac{1}{2}m}}{2^n n!}\frac{\mathrm{d}^{n-m}}{\mathrm{d}\mu^{n-m}}(\mu^2-1)^n. \quad (21.16)$$

In a similar way the solution

$$Q_n^{-m}(\mu)=(-1)^m\frac{\Gamma(\frac{1}{2})\Gamma(n-m+1)}{2^{n+1}\Gamma(n+\frac{3}{2})}\cdot\frac{(\mu^2-1)^{-\frac{1}{2}m}}{\mu^{n-m+1}}$$
$$\times {}_2F_1\left(\tfrac{1}{2}n-\tfrac{1}{2}m+\tfrac{1}{2},\tfrac{1}{2}n-\tfrac{1}{2}m+1;n+\tfrac{3}{2};\frac{1}{\mu^2}\right) \quad (21.17)$$

is derived from equation (21.8) and the result

$$\Gamma(n+1)\int_\infty^\mu\int_\infty^\xi\cdots\int_\infty^\xi\frac{(\frac{1}{2}n+\frac{1}{2})_r(\frac{1}{2}n+1)_r}{\xi^{n+1+2r}}(\mathrm{d}\xi)^m$$
$$=(-1)^m\Gamma(n-m+1)\frac{(\frac{1}{2}n-\frac{1}{2}m+\frac{1}{2})_r(\frac{1}{2}n-\frac{1}{2}m+1)_r}{\mu^{n-m+1+2r}}$$

used in equation (17.7).

The four functions $P_n^m(\mu)$, $Q_n^m(\mu)$, $P_n^{-m}(\mu)$, $Q_n^{-m}(\mu)$ defined in equations (21.9), (21.13), (21.15) and (21.17) respectively are therefore solutions of Legendre's associated equation. They are known as **Legendre's associated functions.** Although the expressions above have been found by assuming m and n to be integers it is readily shown that the solutions quoted are valid even when m and n are not integers. Since Legendre's associated equation is of the second degree it follows that only two of these four functions are linearly independent, and that the other two may be expressed simply in terms of them. It follows immediately from equations (21.11) and (21.15) that if m, n are integers

$$P_n^{-m}(\mu)=\frac{\Gamma(n-m+1)}{\Gamma(n+m+1)}P_n^m(\mu). \quad (21.18)$$

Furthermore, if we apply the result (7.6) to the hypergeometric series on the right-hand side of equation (21.17), we find that

$$Q_n^{-m}(\mu)=(-1)^m\frac{\Gamma(\frac{1}{2})\Gamma(n-m+1)}{2^{n+1}\Gamma(n+\frac{3}{2})}\mu^{-m-n-1}(\mu^2-1)^{\frac{1}{2}m}$$
$$\times {}_2F_1\left(\tfrac{1}{2}m+\tfrac{1}{2}n+\tfrac{1}{2},\tfrac{1}{2}m+\tfrac{1}{2}n+1;n+\tfrac{3}{2};\frac{1}{\mu^2}\right)$$

which, on comparison with equation (21.13), reveals the relation

$$Q_n^{-m}(\mu)=\frac{\Gamma(n-m+1)}{\Gamma(n+m+1)}Q_n^m(\mu). \quad (21.19)$$

It is now a simple matter to prove that when m and n are integers, and m is fixed, the polynomials $P_n^m(\mu)$ form an orthogonal sequence for the interval $[-1, 1]$. Making use of the results (21.18), (21.12) and (21.16) we find that

$$\int_{-1}^{1} P_n^m(\mu)P_{n'}^m(\mu)\,\mathrm{d}\mu = \frac{\Gamma(n+m+1)}{\Gamma(n-m+1)}\frac{1}{2^{n+n'}n!n'!}$$

$$\times\int_{-1}^{1}\frac{\mathrm{d}^{n-m}}{\mathrm{d}\mu^{n-m}}(\mu^2-1)^n\frac{\mathrm{d}^{n'+m}}{\mathrm{d}\mu^{n'+m}}(\mu^2-1)^n\,\mathrm{d}\mu$$

and after integrating by parts $n-m$ times the expression on the right reduces to

$$\frac{\Gamma(n+m+1)}{\Gamma(n-m+1)}\frac{(-1)^{n-m}}{2^{n+n'}n!n'!}\int_{-1}^{1}(\mu^2-1)^n\frac{\mathrm{d}^{n+n'}}{\mathrm{d}\mu^{n+n'}}(\mu^2-1)^{n'}\,\mathrm{d}\mu.$$

This integral is evaluated by the method used at the end of §15 and we find that

$$\int_{-1}^{1} P_n^m(\mu)P_{n'}^m(\mu)\,\mathrm{d}\mu = \frac{\Gamma(n+m+1)}{\Gamma(n-m+1)}(-1)^m\frac{2}{2n+1}\delta_{n,n'} \tag{21.20}$$

In many physical problems $\mu=\cos\theta$, so that $-1\leqslant\mu\leqslant 1$. It is then not always convenient to have a factor of the form $(\mu^2-1)^{\frac{1}{2}m}$. We use instead **Ferrer's function**

$$T_n^m(\mu)=(1-\mu^2)^{\frac{1}{2}m}\frac{\mathrm{d}^m P_n(\mu)}{\mathrm{d}\mu^m}. \tag{21.21}$$

With this notation we may write (21.20) in the form

$$\int_{-1}^{1} T_n^m(\mu)T_{n'}^m(\mu)\,\mathrm{d}\mu = \frac{\Gamma(n+m+1)}{\Gamma(n-m+1)}\frac{2}{2n+1}\delta_{n,n'}. \tag{21.22}$$

The other formulae are amended similarly.

22. Integral expression for the associated Legendre function

From the corollary to Cauchy's theorem

$$f^{(r)}(z)=\frac{r!}{2\pi i}\int_C\frac{f(s)\,\mathrm{d}s}{(s-z)^{r+1}}, \tag{22.1}$$

we deduce from equation (21.1) that

$$P_n^m(\mu)=\frac{(m+n)!\,(\mu^2-1)^{\frac{1}{2}m}}{2^n\,.\,n!}\frac{1}{2\pi i}\int_C\frac{(\zeta^2-1)^n}{(\zeta-\mu)^{m+n+1}}\,d\zeta. \qquad (22.2)$$

If $\mu>0$ we may take the contour C to be the circle

$$|\zeta-\mu|=|\sqrt{(\mu^2-1)}|.$$

Integrating round this contour we obtain from equation (22.2) the equation

$$\frac{1}{2\pi}\int_0^{2\pi}\{\mu+\sqrt{(\mu^2-1)}\cos(\phi-\psi)\}^n\,{\cos \atop \sin}(m\phi)\,d\phi$$
$$=\frac{n!}{(n+m)!}\,{\cos \atop \sin}(m\psi)P_n^m(\mu),$$

from which follows immediately the Fourier expansion

$$\{\mu+\sqrt{(\mu^2-1)}\cos(\phi-\psi)\}^n$$
$$=P_n(\mu)+2\sum_{m=1}^{n}\frac{n!}{(n+m)!}P_n^m(\mu)\cos m(\psi-\phi). \qquad (22.3)$$

Changing n to $-(n+1)$ we obtain the expansion

$$\{\mu'+\sqrt{(\mu'^2-1)}\cos\psi\}^{-n-1}$$
$$=P_n(\mu')+2\sum_{m=1}^{n}(-1)^m\frac{(n-m)!}{n!}P_n^m(\mu')\cos m\psi. \qquad (22.4)$$

Applying Parseval's theorem for Fourier series to the series (22.3) and (22.4) we find that the series

$$P_n(\mu)P_n(\mu')+2\sum_{m=1}^{n}(-1)^m\frac{(n-m)!}{(n+m)!}P_n^m(\mu)P_n^m(\mu')\cos m\phi$$

converges to the sum

$$\frac{1}{2\pi}\int_{-\pi}^{\pi}\frac{\{\mu+\sqrt{(\mu^2-1)}\cos(\psi+\phi)\}^n}{\{\mu'+\sqrt{(\mu'^2-1)}\cos\psi\}^{n+1}}\,d\psi.$$

This integral may be evaluated by means of Cauchy's theorem; for details, see Hobson (1965) pp. 365–371. Its value is found to be

$$P_n\{\mu\mu'+\sqrt{[(\mu^2-1)(\mu'^2-1)]}\cos\phi)\}.$$

Writing $\mu=\cos\theta$, $\mu'=\cos\theta'$ and

$$\cos\Theta=\cos\theta\cos\theta'+\sin\theta\sin\theta'\cos\phi$$

we obtain the result

$$P_n(\cos \Theta) = P_n(\cos \theta)P_n(\cos \theta')$$

$$+2 \sum_{m=1}^{n} (-1)^m \frac{(n-m)!}{(n+m)!} P_n^m(\cos \theta)P_n^m(\cos \theta), \cos (m\phi), \quad (22.5)$$

which is often of value in the solution of problems in wave mechanics.

§23. Surface spherical harmonics

From the two sets of orthogonal functions $T_n^m(\cos \theta)$, $\cos (m\phi)$ we can form a third set of functions

$$X_{n,m}(\theta, \phi) = \left(\frac{2n+1}{2\pi}\right)^{\frac{1}{2}} \left\{\frac{(n-m)!}{(n+m)!}\right\}^{\frac{1}{2}} T_n^m(\cos \theta) \cos m\phi, \qquad m \leqslant n, \qquad (23.1)$$

which is an orthogonal set of functions on the unit sphere, i.e. the functions of the set satisfy the normalisation relation

$$\int_0^{\pi} \sin \theta \, d\theta \int_0^{2\pi} X_{n,m} X_{n',m'} \, d\phi = \delta_{nn'} \delta_{mm'}. \qquad (23.2)$$

In a similar way we can construct a set

$$Y_{n,m}(\theta, \phi) = \left(\frac{2n+1}{2\pi}\right)^{\frac{1}{2}} \left\{\frac{(n-m)!}{(n+m)!}\right\}^{\frac{1}{2}} T_n^m(\cos \theta) \sin m\phi, \qquad m \leqslant n, \qquad (23.3)$$

which satisfies the relations

$$\int_0^{\pi} \sin \theta \, d\theta \int_0^{2\pi} Y_{n,m} Y_{n',m'} \, d\phi = \delta_{nn'} \delta_{mm'}, \qquad (23.4)$$

$$\int_0^{\pi} \sin \theta \, d\theta \int_0^{2\pi} X_{n,m} Y_{n',m'} \, d\phi = 0 \qquad (23.5)$$

for all integral values of n, n', m and m' with $m \leqslant n$, $m' \leqslant n'$.

Because of these orthogonality relationships we can establish an expansion theorem which is a straightforward generalisation of the Legendre series (16.3). It is readily shown that for a large class of functions f, the function $f(\theta, \phi)$ can be represented by the series

$$\sum_{n=0}^{\infty} c_n P_n(\cos\theta) + \sum_{n=1}^{\infty} \sum_{m=1}^{\infty} \{x_{nm} X_{n,m}(\theta, \phi) + y_{nm} Y_{n,m}(\theta, \phi)\}, \tag{23.6}$$

where the coefficients c_n, x_{nm}, y_{nm} are given by the expressions

$$c_n = \frac{2n+1}{4\pi} \int_0^{2\pi} \mathrm{d}\phi \int_0^{\pi} f(\theta, \phi) P_n(\cos\theta) \sin\theta \, \mathrm{d}\theta, \tag{23.7}$$

$$x_{nm} = \int_0^{\pi} \sin\theta \, \mathrm{d}\theta \int_0^{2\pi} X_{n,m}(\theta, \phi) f(\theta, \phi) \, \mathrm{d}\phi, \tag{23.8}$$

$$y_{nm} = \int_0^{\pi} \sin\theta \, \mathrm{d}\theta \int_0^{2\pi} Y_{n,m}(\theta, \phi) f(\theta, \phi) \, \mathrm{d}\phi. \tag{23.9}$$

For any given function $f(\theta, \phi)$ the series (23.6) can therefore in principle be computed by a series of simple integrations.

The functions $X_{n,m}$ and $Y_{n,m}$, which are known as **surface spherical harmonics,** can be constructed easily from the known expressions for the associated functions T_n^m. We find, for instance, that

$$X_{1,1}(\theta, \phi) = -\left(\frac{3}{4\pi}\right)^{\frac{1}{2}} \sin\theta \cos\phi,$$

$$X_{2,1}(\theta, \phi) = -\left(\frac{15}{4\pi}\right)^{\frac{1}{2}} \sin\theta \cos\theta \cos\phi,$$

$$X_{2,2}(\theta, \phi) = \left(\frac{15}{16\pi}\right)^{\frac{1}{2}} \sin^2\theta \cos 2\phi,$$

$$X_{3,1}(\theta, \phi) = -\left(\frac{21}{32\pi}\right)^{\frac{1}{2}} \sin\theta (5\cos^2\theta - 1) \cos\phi,$$

$$X_{3,2}(\theta, \phi) = \left(\frac{105}{16\pi}\right)^{\frac{1}{2}} \sin^2\theta \cos\theta \cos 2\phi,$$

$$X_{3,3}(\theta, \phi) = -\left(\frac{35}{32\pi}\right)^{\frac{1}{2}} \sin^3\theta \cos 3\phi,$$

and the corresponding expressions for the $Y_{n,m}$ are obtained by replacing $\cos(m\phi)$ by $\sin(m\phi)$ in the expressions for the $X_{n,m}$.

The functions $X_{n,m}$ and $Y_{n,m}$ have the important property that they are solutions of the partial differential equation

$$\frac{1}{\sin\theta}\frac{\partial}{\partial\theta}\left(\sin\theta\frac{\partial X}{\partial\theta}\right)+\frac{1}{\sin^2\theta}\frac{\partial^2 X}{\partial\phi^2}+n(n+1)X=0, \qquad (23.10)$$

so that the function

$$(Ar^n+Br^{-n-1})X_{n,m}(\theta,\phi)+(Cr^n+Dr^{-n-1})Y_{n,m}(\theta,\phi),$$

where A, B, C and D are constants, is a solution of Laplace's equation. It follows immediately from equations (23.7)–(23.9) taken with the expansion (23.6) that the function

$$\psi(r,\theta,\phi)=\sum_{n=0}^{\infty}c_n\left(\frac{r}{a}\right)^n P_n(\cos\theta)$$
$$+\sum_{n=1}^{\infty}\sum_{m=1}^{\infty}\left(\frac{r}{a}\right)^n\{x_{nm}X_{n,m}(\theta,\phi)+y_{nm}Y_{n,m}(\theta,\phi)\}$$

satisfies Laplace's equation in the region $0\leqslant r\leqslant a$, is finite at $r=0$, and takes the value $f(\theta,\phi)$ on the sphere $r=a$.

For example, suppose we wish to find the solution of Laplace's equation which takes on the value x^2 on the surface of the sphere $r=a$. Here we have

$$\begin{aligned}f(\theta,\phi)&=a^2\sin^2\theta\cos^2\phi\\&=\frac{a^2}{3}-\frac{a^2}{3}\left(\frac{3\cos^2\theta-1}{2}\right)+\tfrac{1}{2}a^2\sin^2\theta\cos 2\phi\\&=\frac{a^2}{3}-\frac{a^2}{3}P_2(\cos\theta)+\left(\frac{4\pi}{15}\right)^{\frac{1}{2}}X_{2,2}(\theta,\phi)a^2.\end{aligned}$$

Thus the required solution is

$$\psi(r,\theta,\phi)=\tfrac{1}{3}a^2-\tfrac{1}{3}r^2P_2(\cos\theta)+(\tfrac{4}{15})^{\frac{1}{2}}r^2X_{2,2}(\theta,\phi).$$

Substituting the values of P_2 and $X_{2,2}$ and transforming back to Cartesian coordinates we see that the required solution is

$$\psi=\tfrac{1}{3}(a^2+2x^2-y^2-z^2).$$

§24. Use of associated Legendre functions in wave mechanics

To illustrate the use of associated Legendre functions in wave mechanics, we shall consider one of the simplest problems in that subject – that of solving Schrödinger's equation

$$\nabla^2\psi + \frac{8\pi^2 m}{h^2}(W - V)\psi = 0 \tag{24.1}$$

for the rotator with free axis, that is for a particle moving on the surface of a sphere. In equation (24.1), W represents the total energy of the system, V the potential energy. In the case under consideration V is a constant, V_0 say, and the wave function ψ will be a function of θ, ϕ only. If the radius of the sphere is denoted by a, then equation (24.1) is of the form

$$\frac{1}{a^2}\frac{\partial^2\psi}{\partial\theta^2} + \frac{\cot\theta}{a^2}\frac{\partial\psi}{\partial\theta} + \frac{1}{a^2\sin\theta}\frac{\partial^2\psi}{\partial\phi^2} + \frac{8\pi^2 m(W - V_0)}{h^2}\psi = 0. \tag{24.2}$$

If we consider solutions of the form

$$\psi = \Theta e^{\pm im\phi},$$

then

$$\Theta'' + \Theta'\cot\theta + \frac{8\pi^2 ma^2(W - V_0)}{h^2}\Theta - \frac{m^2}{\sin^2\theta}\Theta = 0.$$

Substituting

$$\mu = \cos\theta, \qquad \frac{8\pi^2 ma^2(W - V_0)}{h^2} = n(n+1), \tag{24.3}$$

we find that this equation reduces to Legendre's associated equation (21.1) and hence has solution

$$\Theta = AP_n^m(\cos\theta) + BQ_n^m(\cos\theta).$$

However, for the same reason as in the case of potential theory (§20 above), we must take $B = 0$. The solutions of equation (24.2) will therefore be made up of combinations of solutions of the form

$$\psi_{m,n}(\theta, \phi) = A_{mn}e^{\pm im\phi}P_n^m(\cos\theta), \tag{24.4}$$

where A_{mn} is a constant.

The physical conditions imposed on the wave function ψ are that it should be single-valued and continuous. Obviously then the 'physical' solutions will have m an integer, since $\psi_{m,n}(\theta, \phi + 2\pi)$ must equal $\psi_{m,n}(\theta, \phi)$. Further, in order that the series for $P_n^m(\mu)$ should converge for the values $\mu = \pm 1$ it is necessary that it should have only a finite number of terms. This is possible only if n is a positive integer. If therefore the solution (24.4) is to be valid for $\theta = 0$ and $\theta = \pi$, we must have n a positive integer. The physical conditions on the wave function are therefore not satisfied by systems with an arbitrary value for the energy W, but only by systems for which

$$W = V_0 + \frac{h^2}{8\pi^2 ma^2} n(n+1), \tag{24.5}$$

where n is a positive integer. In other words, the energy of such a mechanical system does not vary continuously, but is capable of assuming values taken from the discrete set (24.5).

Problems III

3.1 Show that, if n is a positive integer,

$$P_{2n}(0) = (-1)^n \frac{(2n)!}{2^{2n}(n!)^2}, \qquad P_{2n+1}(0) = 0,$$

$$P'_{2n}(0) = 0, \qquad P'_{2n+1}(0) = (-1)^n \frac{(2n+1)!}{2^{2n}(n!)^2}$$

$$P''_{2n}(0) = (-1)^{-1} \frac{(2n+1)!}{2^{2n-1} n!(n-1)!}, \qquad P''_{2n+1}(0) = 0.$$

3.2 Prove that

$$\sum_{n=0}^{\infty} \frac{\mu^{n+1}}{n+1} P_n(\mu) = \tfrac{1}{2} \log \left\{ \frac{1+\mu}{1-\mu} \right\}.$$

3.3 Given that $\zeta = \mu + \sqrt{(\mu^2 - 1)}$ show that

(i) $(1 - h\zeta)^{-\frac{1}{2}}(1 - h/\zeta)^{-\frac{1}{2}} = \sum_{n=0}^{\infty} h^n P_n(\mu)$;

(ii) $P_n(\mu) = \dfrac{\Gamma(n+\frac{1}{2})}{n!\Gamma(\frac{1}{2})} \zeta^n \, {}_2F_1(\frac{1}{2}, -n; \frac{1}{2} - n; \zeta^{-2})$.

Deduce that

(iii) ${}_2F_1(\frac{1}{2}, -n; \frac{1}{2} - n; 1) = \dfrac{n!\Gamma(\frac{1}{2})}{\Gamma(n+\frac{1}{2})}$.

3.4 Prove that if n is a positive integer,

$$\int_{-1}^{1} P_n(\mu)(1 - 2\mu h + h^2)^{-\frac{1}{2}} \, d\mu = \frac{2h^n}{2n+1},$$

and hence, making use of Rodrigues' formula, deduce that

$$\int_{-1}^{1} (1-\mu^2)^n(1-2\mu h+h^2)^{-n-\frac{1}{2}}\,d\mu = \frac{2^{2n+1}(n!)^2}{(2n+1)!}.$$

3.5 Prove that

$$P_n'(x) = \sum_{r=0}^{p} (2n-4r-1)P_{n-2r-1}(x),$$

where $p=\frac{1}{2}(n-1)$ or $\frac{1}{2}n-1$ according as n is odd or even.

Deduce that for all x in the closed interval $[-1, 1]$ and for all positive integers n, the values of the functions

$$|P_n(x)|, \quad n^{-2}\,|P_n'(x)|, \quad n^{-4}\,|P''(x)|, \ldots$$

can never exceed unity.

3.6 Prove that

$$\int_{\mu}^{1} P_n(\mu)\,d\mu = \frac{1}{2n+1}\{P_{n-1}(\mu)-P_{n+1}(\mu)\},$$

and deduce, from Problem 3.1, that if n is an odd integer

$$\int_{0}^{1} P_n(\mu)\,d\mu = \frac{(-1)^{\frac{1}{2}n-\frac{1}{2}}(n-1)!}{2^n(\frac{1}{2}n+\frac{1}{2})!(\frac{1}{2}n-\frac{1}{2})!}.$$

What is the value of the integral when n is even?

3.7 Using equation (14.2) and the results of the last example, show that, if n is even,

$$\int_{0}^{1} \mu P_n(\mu)\,d\mu = (-1)^{\frac{1}{2}n-1}\frac{(n-2)!}{2^n(\frac{1}{2}n+1)!(\frac{1}{2}n-1)!},$$

and that the integral has the value zero if n is odd.

3.8 Given that

$$u_n = \int_{-1}^{1} P_n(\mu)P_{n-1}(\mu)\frac{d\mu}{\mu},$$

prove that $(n+1)u_{n+1}+nu_n=2$. Hence evaluate u_n.

3.9 Prove that if m and n are positive integers,

$$\int_{-1}^{1} (1+\mu)^{m+n}P_n(\mu)\,d\mu = \frac{2^{m+n+1}\{(m+n)!\}^2}{m!(m+2n+1)!}.$$

3.10 Prove that if n is even and $m>-1$,

$$\int_0^1 \mu^m P_n(\mu)\,\mathrm{d}\mu = \frac{\Gamma(\frac{1}{2}m+\frac{1}{2})\Gamma(\frac{1}{2}m+1)}{2\Gamma(\frac{1}{2}m+\frac{1}{2}n+\frac{3}{2})\Gamma(\frac{1}{2}m-\frac{1}{2}n+1)}.$$

Deduce that

$$\int_0^1 (1-k\mu^2)^{-n-\frac{3}{2}} P_{2n}(\mu)\,\mathrm{d}\mu = \frac{1}{2n+1}k^n(1-k)^{-n-\frac{1}{2}}.$$

3.11 Show that

$$P_n(\mu) = \left(\frac{\mu+1}{2}\right)^n {}_2F_1\left(-n;-n,1;\frac{\mu-1}{\mu+1}\right),$$

and hence that

$$(1-t)^n P_n\left(\frac{1+t}{1-t}\right) = \sum_{r=0}^{n} ({}^nC_r)^2 t^r.$$

Deduce that

$$P_n(\cosh u) \geqslant 1.$$

3.12 Given that $f(\mu)=(\mu^2-1)^n$ show, by using Rolle's theorems that $f'(\mu)$ must have at least one zero between -1 and 1. Proceeding in this way deduce that $f^{(n)}(\mu)$ has n zero, between -1 and 1.

Hence show that when n is even the zeros of $P_n(\mu)$ occur in pairs, equal in magnitude but opposite in sign, and that when n is odd, $\mu=0$ is a zero and the others occur in equal and opposite pairs.

3.13 Let $y(\mu)$ be any solution of the linear differential equation

$$\alpha(\mu)\frac{\mathrm{d}^2y}{\mathrm{d}\mu^2}+\beta(\mu)\frac{\mathrm{d}y}{\mathrm{d}\mu}+\gamma(\mu)y=0$$

in which α, β and γ are continuous functions of μ whose derivatives of all orders are continuous. Prove that $y(\mu)$ cannot have any repeated zeros except possibly for values of μ which satisfy the equation $\alpha(\mu)=0$.

Deduce that all the zeros of $P_n(\mu)$ are distinct.

3.14 Prove that the Legendre polynomial $P_n(x)$ has the smallest distance in the mean from zero of all polynomials of degree n with leading coefficient $2^n(\frac{1}{2})_n/n!$

3.15 Prove that if R denotes the operator

$$\frac{\mathrm{d}}{\mathrm{d}x}\left\{(1-x^2)\frac{\mathrm{d}}{\mathrm{d}x}\right\},$$

then

$$\int_{-1}^{1} P_n(x)R\{f(x)\}\,\mathrm{d}x = -n(n+1)\int_{-1}^{1} P_n(x)f(x)\,\mathrm{d}x$$

provided that $f(x)$ and $f'(x)$ are finite at $x=\pm 1$.

Prove that if $n \geqslant 1$,

$$\int_{-1}^{1} \log(1-x)P_n(x)\,\mathrm{d}x = -\frac{2}{n(n+1)}.$$

3.16 Prove that

(i) $$\int_{-1}^{1} \frac{P_n(x)\,\mathrm{d}x}{\surd(1-x)} = \frac{2\surd 2}{2n+1};$$

(ii) $$\int_{-1}^{1} \frac{P_n(x)\,\mathrm{d}x}{(1-2hx+h^2)^{\frac{3}{2}}} = \frac{2h^n}{1-h^2}.$$

3.17 Prove that if $R^2 = 1-2hx+h^2$, and $|h|<1$, $n \geqslant 1$,

(i) $$\int_{-1}^{1} \log\left\{\frac{h-x+R}{1-x}\right\}P_n(x)\,\mathrm{d}x = \frac{2h^{n+1}}{(2n+1)(n+1)};$$

(ii) $$\int_{-1}^{1} \log(1-hx+R)P_n(x)\,\mathrm{d}x = -\frac{2h^n}{n(2n+1)}.$$

3.18 Given that

$$f(x, a) = \int_0^a \frac{\xi^{m-1}}{(1-2\xi x+\xi^2)^{\frac{1}{2}}}\,\mathrm{d}\xi, \qquad m>1,$$

prove that

$$\int_{-1}^{1} f(x, a)P_n(x)\,\mathrm{d}x = \frac{2a^{n+m}}{(n+m)(2n+1)}.$$

3.19 Prove that if z is real and $|z|<1$,

$$\int_0^{\pi} \frac{\mathrm{d}\phi}{1+z\cos\phi} = \frac{\pi}{\surd(1-z^2)}.$$

Putting $z = \mp h\surd(\mu^2-1)/(1-h\mu)$ where h is so small that

$$|h\{\mu \pm \surd(\mu^2-1)\cos\phi\}|<1$$

($0 \leqslant \phi \leqslant \pi$), expanding both sides in powers of h, and equating coefficients of h^n, show that

$$P_n(\mu) = \frac{1}{\pi} \int_0^{\pi} \{\mu \pm \sqrt{(\mu^2 - 1)} \cos \phi\}^n \, d\phi.$$

Hence evaluate the sum

$$\sum_{r=0}^{n} {}^nC_r P_r(\cos \theta).$$

3.20 Show by making the substitution

$$z = \pm h\sqrt{(\mu^2 - 1)}/(h\mu - 1)$$

in the formula in Problem 3.19, that

$$P_n(\mu) = \frac{1}{\pi} \int_0^{\pi} \frac{d\phi}{\{\mu + \sqrt{(\mu^2 - 1)} \cos \phi\}^{n+1}}.$$

3.21 Making use of the integral expression for $P_n(\mu)$ derived in Problem 3.19, show that

$$P_n(\mu) = \sum_{r=0}^{n} {}^nC_r (1 - \mu^2)^{\frac{1}{2}n - \frac{1}{2}r} \mu^r P_{n-r}(0).$$

Deduce that

$$\begin{vmatrix} P_0(\mu) & P_1(\mu) & P_2(\mu) \\ P_1(\mu) & P_2(\mu) & P_3(\mu) \\ P_2(\mu) & P_3(\mu) & P_4(\mu) \end{vmatrix} = (1 - \mu^2)^3 \begin{vmatrix} P_0(0) & 0 & P_2(0) \\ 0 & P_2(0) & 0 \\ P_2(0) & 0 & P_4(0) \end{vmatrix}.$$

3.22 Show that

$$\sum_{j=0}^{n} (2j+1) P_j(x) P_j(y) = (n+1) \frac{P_{n+1}(x) P_n(y) - P_n(x) P_{n+1}(y)}{x - y},$$

and deduce that if $\{x_k\}_{k=1}^n$ are the zeros of $P_n(x)$, then

$$\int_{-1}^{1} \frac{P_n(x) \, dx}{x - x_k} = \frac{2}{n P_{n-1}(x_k)}.$$

Hence derive the formula

$$\int_{-1}^{1} f(x) \, dx = \sum_{k=1}^{n} w_k f(x_k) + R_n(f),$$

where the weights $\{w_k\}_{k=1}^n$ are defined in terms of the zeros

$\{x_k\}_{k=1}$ of $P_n(x)$ by the equation

$$w_k = \frac{2}{nP_n'(x_k)P_{n-1}(x_k)}$$

and $R_n(f)$ is defined by equation (4.19).

3.23 Prove that if $\mu > 1$

$$Q_n(\mu) = \frac{1}{2^{n+1}} \int_{-1}^{1} \frac{(1-t^2)^n}{(\mu-t)^{n+1}}\, dt,$$

and deduce that

$$Q_n(\mu) = \int_0^{\alpha} \{\mu - \sqrt{(\mu^2-1)} \cosh \theta\}^n \, d\theta,$$

where $\alpha = \frac{1}{2} \log (\mu+1)/(\mu-1)$.

Hence find expressions for $Q_0(\mu)$ and $Q_1(\mu)$.

3.24 Determine the simple expressions for $Q_0(\mu)$, $Q_1(\beta)$, $Q_2(\mu)$ and $Q_3(\mu)$ by working out the values of the polynomial $W_{n-1}(\mu)$, occurring in equation (18.8), for these values of n.

3.25 Establish the following formulae due to MacRobert:

(i) $$\mu P_m(\mu) Q_n(\mu) = \frac{1}{2} \int_{-1}^{1} \frac{\xi P_m(\xi) P_n(\xi)}{\mu - \xi}\, d\xi, \quad m < n;$$

(ii) $$\mu P_n(\mu) Q_n(\mu) = \frac{1}{2} \int_{-1}^{1} \frac{\xi \{P_n(\xi)\}^2}{\mu - \xi}\, d\xi + \frac{1}{2n+1};$$

(iii) $$\mu^m (\mu^2-1) Q_n'(\mu) = \frac{1}{2} \int_{-1}^{1} \frac{\xi^m (\xi^2-1) P_n'(\xi)}{\mu - \xi}\, d\xi, \quad n > m;$$

(iv) $$(\mu^2-1) P_m(\mu) Q_n'(\mu) = \frac{1}{2} \int_{-1}^{1} \frac{(\xi^2-1) P_m(\xi) P_n'(\xi)}{\mu - \xi}\, d\xi, \quad n > m.$$

3.26 Prove that, if n is a positive integer and $\zeta = \mu + \sqrt{(\mu^2-1)}$,

$$Q_n(\mu) = \zeta \int_0^{\pi} \frac{P_n(\cos \theta) \sin \theta \, d\theta}{1 - 2\zeta \cos \theta + \zeta^2}.$$

Deduce that, if $|\zeta| > 1$,

$$Q_n(\mu) = \sum_{m=1}^{\infty} \zeta^{-m} \int_0^{\pi} P_n(\cos \theta) \sin m\theta \, d\theta.$$

By evaluating this integral show that

$$Q_n(\mu)=\frac{\sqrt{\pi}n!}{\Gamma(n+\frac{3}{2})}\zeta^{-n-1}\,{}_2F_1(\tfrac{1}{2},n+1;n+\tfrac{3}{2};\zeta^{-2}).$$

3.27 Prove that, if m is a positive integer,

$$\sum_{n=m}^{\infty} h^{n-m}T_n^m(\mu)=\frac{(-1)^m(2m)!\,(1-\mu^2)^{\frac{1}{2}m}}{2^m m!\,(1-2\mu h+h^2)^{m+\frac{1}{2}}}.$$

3.28 Show that, if $|\mu|>1$, $n>1$, $n>-1$ then

$$Q_n^m(\mu)=\frac{\Gamma(n+m+1)}{\Gamma(n+1)}\frac{(\mu^2-1)^{\frac{1}{2}m}}{2^{n+1}}\int_{-1}^{1}\frac{(1-\xi^2)^n\,d\xi}{(\mu-\xi)^{n+m+1}}.$$

Deduce that, if $|\mu+1|>2$ then

$$Q_n^m(\mu)=\frac{\sqrt{\pi}\Gamma(n+m+1)}{2^{n-1}\Gamma(n+\frac{3}{2})}\frac{(\mu^2-1)^{\frac{1}{2}m}}{(\mu+1)^{n+m+1}}$$

$$\times{}_2F_1\left(n+2,n+m+1;2n+2;\frac{2}{\mu+1}\right).$$

Find a simple expression for $Q_n^{n+1}(\mu)$.

3.29 Prove the following recurrence relations for Ferrer's Associated Legendre Functions:

(i) $T_{n+1}^{m+1}(\mu)-T_{n-1}^{m+1}(\mu)=(2n+1)(1-\mu^2)^{\frac{1}{2}}T_n^m(\mu)$;

(ii) $(n-m+1)T_{n+1}^m(\mu)-(2n+1)\mu T_n^m(\mu)+(n+m)T_{n-1}^m(\mu)=0$;

(iii) $T_n^{m+1}(\mu)-T_{n-1}^{m+1}(\mu)=(n-m)(1-\mu^2)^{\frac{1}{2}}T_n^m(\mu)$.

3.30 Derive the expressions for $T_4^m(\theta,\phi)$ and $X_4^m(\theta,\phi)$ for $m=$ 1, 2, 3, 4. Express the functions $\sin^2\theta\sin^2\phi$, $\sin^2\theta\cos^2\phi$, $\sin\theta\cos^3\theta\cos\phi$ in terms of surface spherical harmonics.

3.31 Find the function which satisfies Laplace's equation in the interior of the sphere $x^2+y^2+z^2=a^2$, remains finite at the origin and takes the value $\alpha x^2+\beta y^2+\gamma z^2$ on the surface of the sphere.

3.32 The *Jacobi polynomials* are defined by the equation†

$$\mathscr{F}_m(a,b,x)={}_2F_1(-m,a+m;b;x).$$

† It should be observed that this name is sometimes applied to the polynomial

$$P_n^{(\alpha,\beta)}(x)=\frac{(a+1)_n}{n!}\mathscr{F}_n(\alpha+\beta+1,\alpha+1,\tfrac{1}{2}-\tfrac{1}{2}x).$$

Show that if $D \equiv \mathrm{d}/\mathrm{d}x$,

$$D^m\{x^{b+m-1}(1-x)^{a+m-b}\} = (b)_m x^{b-1}\,{}_2F_1(b-a-m, b+m; b; x)$$

Deduce that

(i) $\mathscr{F}_m(a, b, x) = \dfrac{x^{1-b}(1-x)^{b-a}}{(b)_m} D^m\{x^{b+m-1}(1-x)^{a+m-b}\};$

(ii) $\displaystyle\int_0^1 x^{b-1}(1-x)^{a-b} f(x)\mathscr{F}_m(a, b, x)\,\mathrm{d}x$

$$= \frac{1}{(b)_m}\int_0^1 (-1)^m f^{(m)}(x) x^{b+m-1}(1-x)^{a+m-b}\,\mathrm{d}x;$$

(iii) $\displaystyle\int_0^1 x^{b-1}(1-x)^{a-b}\mathscr{F}_m(a, b, x)\mathscr{F}_n(a, b, x)\,\mathrm{d}x = 0, \qquad m \neq n;$

(iv) $\displaystyle\int_0^1 x^{b-1}(1-x)^{a-b}\{\mathscr{F}_m(a, b, x)\}^2$

$$= \frac{\Gamma(b)\Gamma(a+1-b)(a+1-b)_m}{\Gamma(a)(a)_m(b)_m}\,\frac{m!}{a+2m}$$

3.33 The Chebyshev polynomials of the first and second kinds are defined by the equations

$$T_n(\cos\theta) = \cos(n\theta), \qquad U_n(\cos\theta) = \frac{\sin(n+1)\theta}{\sin\theta}.$$

Making use of Problem 2.2, show that

$$T_n(x) = {}_2F_1\left(-n, n; \tfrac{1}{2}; \frac{1-x}{2}\right),$$

$$U_n(x) = (n+1)\,{}_2F_1\left(-n, n+1; \tfrac{3}{2}; \frac{1-x}{2}\right).$$

Prove also that

(i) $1 + 2\displaystyle\sum_{n=1}^{\infty} T_n(x)t^n = (1-t^2)(1-2tx+t^2)^{-1}, \quad |t|<1, \quad |x|<1;$

(ii) $2^n(\tfrac{1}{2})_n T_n(x) = (-1)^n(1-x^2)^{\frac{1}{2}n}\dfrac{\mathrm{d}^n}{\mathrm{d}x^n}(1-x^2)^{n-\frac{1}{2}};$

(iii) $\displaystyle\int_{-1}^1 T_m(x)T_n(x)(1-x^2)^{-\frac{1}{2}}\,\mathrm{d}x = \tfrac{1}{2}\pi\delta_{mn};$

(iv) $\sum_{n=0}^{\infty} U_n(x)t^n = (1-2xt+t^2)^{-1}$;

(v) $2^{n+1}(\frac{1}{2})_{n+1} U_n(x) = (-1)^n (n+1)(1-x^2)^{-\frac{1}{2}} \frac{d^n}{dx^n}(1-x^2)^{n+\frac{1}{2}}$;

(vi) $\int_{-1}^{1} U_m(x)U_n(x)(1-x^2)^{\frac{1}{2}}\,dx = \frac{1}{2}\pi\delta_{mn}$;

(vii) $T_n(x) = U_n(x) - xU_{n-1}(x)$;

(viii) $(1-x^2)U_{n-1}(x) = xT_n(x) - T_{n+1}(x)$.

3.34 Prove that

$$1+2\sum_{j=1}^{N} T_j(x)T_j(y) = \frac{T_{n+1}(x)T_n(y) - T_n(x)T_{n+1}(y)}{x-y}$$

and deduce that

$$\int_{-1}^{1} \frac{T_n(x)\,dx}{(x-x_k)\sqrt{(1-x^2)}} = \frac{\pi}{T_{n-1}(x_k)},$$

where $\{x_k\}_{k=1}^{n}$ are the zeros of $T_n(x)$, so that

$$x_k = \cos\left(\frac{2k-1}{2n}\pi\right).$$

Hence derive the formula

$$\int_{-1}^{1} \frac{f(x)\,dx}{\sqrt{(1-x)^2}} = \frac{\pi}{n}\sum_{k=1}^{n} f(x_k) + R_n(f),$$

where $R_n(f)$ is defined by equation (4.19).

3.35 Show that

$$2\sum_{j=0}^{n} U_j(x)U_j(y) = \frac{U_{n+1}(x)U_n(y) - U_n(x)U_{n+1}(y)}{x-y}$$

and deduce that

$$\int_{-1}^{1} \sqrt{(1-x^2)}\frac{U_n(x)\,dx}{x-y_k} = \frac{\pi}{U_{n-1}(y_k)},$$

where $\{y_k\}_{k=1}^{n}$ are the zeros of $U_n(x)$, so that

$$y_k = \cos\left(\frac{k}{n+1}\pi\right).$$

Hence derive the formula

$$\int_{-1}^{1} \sqrt{(1-x^2)} f(x)\,dx = \frac{\pi}{n+1} \sum_{k=1}^{n} (1-y_k^2)^{\frac{1}{2}} f(y_k) + R_n(f),$$

where $R_n(f)$ is defined as before.

3.36 The function $C_n^\nu(x)$ defined by

$$(1-2xt+t^2)^{-\nu} = \sum_{n=0}^{\infty} C_n^\nu(x) t^n, \qquad \nu > 0,$$

is called the *Gegenbauer polynomial* of degree n and order ν

Prove the relations:

(i) $C_n^\nu(x) = \dfrac{\Gamma(n+2\nu)}{n!\Gamma(2\nu)}\, {}_2F_1\left(n+2\nu, -n; \nu+\frac{1}{2}; \dfrac{1-x}{2}\right);$

(ii) $C_n^\nu(1) = (-1)^n c_n^\nu(-1) = \dfrac{\Gamma(n+2\nu)}{n!\Gamma(2\nu)};$

(iii) $C_n^\nu(x) = (-2)^n \dfrac{\Gamma(\nu+n)\Gamma(2\nu+n)}{n!\,\Gamma(\nu)\Gamma(2\nu+2n)} (1-x^2)^{\frac{1}{2}-\nu} \dfrac{d^n}{dx^n} (1-x^2)^{n+\nu-\frac{1}{2}};$

(iv) $\displaystyle\int_{-1}^{1} C_n^\nu(x) C_m^\nu(x) (1-x^2)^{\nu-\frac{1}{2}}\,dx = \frac{2^{1-2\nu}\pi\Gamma(n+2\nu)}{n!(\nu+n)[\Gamma(\nu)]^2}\,\delta_{mn}.$

4

Bessel functions

§25. The origin of Bessel functions

Bessel functions were first introduced by Bessel, in 1824, in the discussion of a problem in dynamical astronomy, which may be described as follows. If P is a planet moving in an ellipse whose focus S is the sun and whose centre and major axis are C and $A'A$ respectively (cf. Fig. 8), then the angle ASP is called the *true anomaly* of the planet. It is found that, in astronomical calculations, the true anomaly is not a very convenient angle with which to deal. Instead we use the *mean anomaly* ζ, which is defined to be 2π times the ratio of the area of the elliptic sector ASP to the area of the ellipse. Another angle of significance is the *eccentric anomaly*, u, of the planet, defined to be the angle ACQ, where Q is the point in which the ordinate through P meets the auxiliary circle of the ellipse.

It is readily shown by a simple geometrical argument that, if e is the eccentricity of the ellipse, the relation between the mean anomaly and the eccentric anomaly is

$$\zeta = u - e \sin u. \tag{25.1}$$

The problem set by Bessel was that of expressing the difference between the mean and eccentric anomalies, $u-\zeta$, as a series of sines of multiples of the mean anomaly, i.e. that of determining the coefficients c_r $(r=1, 2, 3, \ldots)$ such that

$$u - \zeta = \sum_{r=1}^{\infty} c_r \sin(r\zeta). \tag{25.2}$$

To obtain the values of the coefficients c_r we multiply both sides of equation (25.2) by $\sin(s\zeta)$ and integrate with respect to ζ from 0 to π. We then obtain

$$\int_0^{\pi} (u-\zeta)\sin(s\zeta)\,d\zeta = \sum_{r=1}^{\infty} c_r \int_0^{\pi} \sin(r\zeta)\sin(s\zeta)\,d\zeta.$$

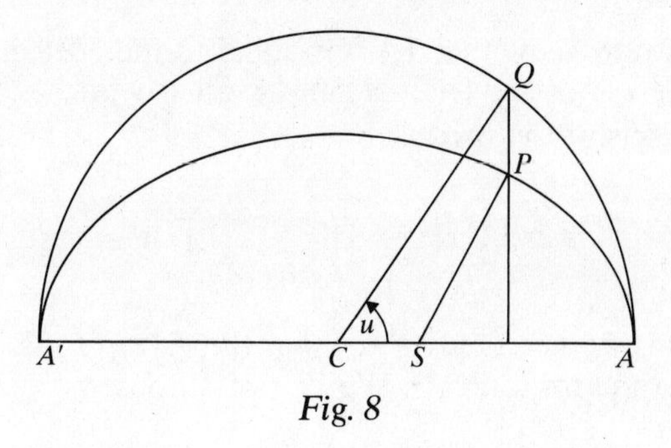

Fig. 8

Now

$$\int_0^\pi \sin(r\zeta)\sin(s\zeta)\,\mathrm{d}\zeta = \tfrac{1}{2}\pi\delta_{r,s}$$

and an integration by parts shows that

$$\int_0^\pi (u-\zeta)\sin(s\zeta)\,\mathrm{d}\zeta = \frac{1}{s}\Big[(\zeta-u)\cos(s\zeta)\Big]_0^\pi + \frac{1}{s}\int_0^\pi \left(\frac{\mathrm{d}u}{\mathrm{d}\zeta}-1\right)\cos(s\zeta)\,\mathrm{d}\zeta.$$

From (25.2), $\zeta - u$ is zero when $\zeta = 0$ and when $\zeta = \pi$, so that the square bracket vanishes; the integral can be written in the form

$$\frac{1}{s}\int_0^\pi \cos s\zeta\,\mathrm{d}u$$

and hence, using equations (25.1) we obtain the result

$$c_s = \frac{2}{\pi s}\int_0^\pi \cos\{s(u - e\sin u)\}\,\mathrm{d}u. \tag{25.3}$$

The integral on the right-hand side of equation (25.3) is a function of s and of the eccentricity e of the planet's orbit. If we write

$$J_n(x) = \frac{1}{\pi}\int_0^\pi \cos(x\sin\theta - n\theta)\,\mathrm{d}\theta \tag{25.4}$$

it follows from equations (25.3) and (25.2) that

$$u - \zeta = 2\sum_{r=1}^{\infty} J_r(er)\frac{\sin(r\zeta)}{r}. \tag{25.5}$$

The function $J_n(x)$ so defined is called **Bessel's coefficient of order *n*.**

We shall now show that $J_n(x)$ is equal to the coefficient of t^n in the expansion of $\exp\{\frac{1}{2}x(t-t^{-1})\}$; in other words we may define $J_n(x)$ by means of the expansion

$$\exp\left\{\tfrac{1}{2}x\left(t-\frac{1}{t}\right)\right\}=\sum_{n=-\infty}^{\infty} J_n(x)t^n. \tag{25.6}$$

To prove this we need only show that the $J_n(x)$ of (25.6) can be expressed in the form (25.4). We first of all observe that

$$\sum_{n=-\infty}^{\infty}(-1)^n J_{-n}(x)t^n=\sum_{n=-\infty}^{\infty} J_n(x)\left(-\frac{1}{t}\right)^n=\sum_{n=-\infty}^{\infty} J_n(x)t^n,$$

since both expansions are equal to $\exp\{\frac{1}{2}x(t-1)\}$. Equating coefficients of t^n we have

$$(-1)^n J_{-n}(x)=J_n(x). \tag{25.7}$$

In the expansion (25.6) we may write $t=e^{i\theta}$ to obtain the relation

$$\exp(ix\sin\theta)=\sum_{n=-\infty}^{\infty} J_n(x)e^{in\theta}.$$

Making use of the result (25.7) we see that the series on the right can be put into the form

$$J_0(x)+2\sum_{m=1}^{\infty} J_{2m}(x)\cos(2m\theta)+2i\sum_{m=0}^{\infty} J_{2m+1}(x)\sin(2m+1)\theta,$$

so that by equating real and imaginary parts we obtain the expansions

$$\cos(x\sin\theta)=J_0(x)+2\sum_{m=1}^{\infty} J_{2m}(x)\cos(2m\theta), \tag{25.8}$$

$$\sin(x\sin\theta)=2\sum_{m=0}^{\infty} J_{2m+1}(x)\sin(2m+1)\theta. \tag{25.9}$$

If we now multiply (25.8) by $\cos n\theta$, (25.9) by $\sin n\theta$, integrate with respect to θ from 0 to π, and use the formulae

$$\int_0^{\pi}\cos(m\theta)\cos(n\theta)\,d\theta=\int_0^{\pi}\sin(m\theta)\sin(n\theta)\,d\theta=\frac{\pi}{2}\,\delta_{m,n},$$

we obtain the formulae

$$J_n(x)=\frac{1}{\pi}\int_0^{\pi} \cos(x \sin\theta)\cos(n\theta)\,d\theta, \qquad n \text{ even}, \qquad (25.10)$$

$$J_n(x)=\frac{1}{\pi}\int_0^{\pi} \sin(x \sin\theta)\sin(n\theta)\,d\theta, \qquad n \text{ odd}. \qquad (25.11)$$

Because of the periodic properties of the trigonometric functions we know that the integral on the right of equation (25.10) is zero if n is odd, while that on the right of equation (25.11) is zero if n is even. Thus for all integral values of n, we have

$$J_n(x)=\frac{1}{\pi}\int_0^{\pi} \{\cos(x \sin\theta)\cos(n\theta)+\sin(x\sin\theta)\sin(n\theta)\}\,d\theta,$$

which is identical with the expression (25.4).

In particular

$$J_0(x)=\frac{1}{\pi}\int_0^{\pi} \cos(x \sin\theta)\,d\theta. \qquad (25.12)$$

In what follows we shall assume that the **Bessel coefficients** of the first kind are defined by equation (25.6) or, which is equivalent, by equation (25.4).

§26. Recurrence relations for the Bessel coefficients

If we differentiate the generating equation (25.6) with respect to x we obtain the relation

$$\frac{1}{2}\left(t-\frac{1}{t}\right)\exp\left\{\tfrac{1}{2}x\left(t-\frac{1}{t}\right)\right\}=\sum_{n=-\infty}^{\infty} J_n'(x)t^n,$$

which is equivalent to

$$\tfrac{1}{2}\sum_{n=-\infty}^{\infty}\{J_n(x)t^{n+1}-J_n(x)t^{n-1}\}-\sum_{n=-\infty}^{\infty} J_n'(x)t^n=0.$$

Equating to zero the coefficient of t^n we obtain the relation

$$2J_n'(x)=J_{n-1}(x)-J_{n+1}(x). \qquad (26.1)$$

On the other hand, if we differentiate (25.6) with respect to t

the resulting equation is

$$\tfrac{1}{2}x\left(1+\frac{1}{t^2}\right)\exp\left\{\tfrac{1}{2}x\left(t-\frac{1}{t}\right)\right\}=\sum_{n=-\infty}^{\infty} nJ_n(x)t^{n-1}$$

and this is equivalent to the relation

$$\tfrac{1}{2}x\sum_{n=-\infty}^{\infty}(t^n+t^{n-2})J_n(x)-\sum_{n=-\infty}^{\infty} nJ_n(x)t^{n-1}=0.$$

Equating the coefficient of t^{n-1} to zero we obtain the recurrence relation

$$\frac{2n}{x}J_n(x)=J_{n-1}(x)+J_{n+1}(x). \tag{26.2}$$

Adding equations (26.1) and (26.2) we find that

$$xJ_n'(x)=xJ_{n-1}(x)-nJ_n(x) \tag{26.3}$$

and subtracting equation (26.1) from (26.2) we obtain

$$xJ_n'(x)=nJ_n(x)-xJ_{n+1}(x). \tag{26.4}$$

Putting $n=0$ in this last equation we have the important special case

$$J_0'(x)=-J_1(x), \tag{26.5}$$

and putting $n=1$ in equation (26.3) we find that

$$J_1'(x)=J_0(x)-\frac{1}{x}J_1(x). \tag{26.6}$$

Differentiating both sides of (26.5) with respect to x and making use of the result (26.6) we have

$$J_0''(x)=-J_0(x)+\frac{1}{x}J_1(x)$$

which, as a consequence of equation (26.6), may be written

$$J_0''(x)+\frac{1}{x}J_0'(x)+J_0(x)=0 \tag{26.7}$$

showing that $y=J_0(x)$ is a solution of the differential equation

$$\frac{d^2y}{dx^2}+\frac{1}{x}\frac{dy}{dx}+y=0. \tag{26.8}$$

We can show similarly that the Bessel function $J_n(x)$ satisfies the differential equation

$$\frac{d^2y}{dx^2}+\frac{1}{x}\frac{dy}{dx}+\left(1-\frac{n^2}{x^2}\right)y=0, \qquad n \text{ integral.} \tag{26.9}$$

For, from equation (26.4) we find, as a result of differentiating both sides with respect to x, that

$$xJ_n''(x)+J_n'(x)=nJ_n'(x)-J_{n+1}(x)-xJ_{n+1}'(x).$$

Now, from equation (26.4),

$$nJ_n'(x)=\frac{n^2}{x^2}J_n(x)-nJ_{n+1}(x),$$

and putting $n+1$ in place of n in equation (26.3) we see that

$$xJ_{n+1}'(x)+(n+1)J_{n+1}(x)=xJ_n(x),$$

so that

$$xJ_n''(x)+J_n'(x)=\frac{n^2}{x}J_n(x)-xJ_n(x),$$

which shows that $J_n(x)$ is a solution of equation (26.9) provided, of course, that n is an integer.

As we pointed out in §1, equation (26.9) is known as **Bessel's equation.** What we have shown is that if the n which occurs in Bessel's equation is an integer, one solution of the equation is $J_n(x)$. It is because of this fact that Bessel coefficients are of such importance in mathematical physics, for as we saw in §1, the equation (26.9) arises naturally in boundary value problems in mathematical physics.

§27. Series expansion for the Bessel coefficients

We shall now find the power series expansion for the Bessel coefficient $J_n(x)$. If we write

$$\exp\left\{\tfrac{1}{2}x\left(t-\frac{1}{t}\right)\right\}=\exp\left(\tfrac{1}{2}xt\right)\exp\left(-\frac{x}{2t}\right)$$

and make use of the power series for the exponential function, we obtain the expansion

$$\exp\left\{\tfrac{1}{2}x\left(t-\frac{1}{t}\right)\right\}=\sum_{r=0}^{\infty}\frac{(xt)^r}{2^r r!}\sum_{s=0}^{\infty}\frac{(-x)^s}{2^s t^s s!}$$

$$=\sum_{r=0}^{\infty}\sum_{s=0}^{\infty}(-1)^s\left(\frac{x}{2}\right)^{r+s}\frac{t^{r-s}}{r!s!}. \tag{27.1}$$

By our definition (25.6), the Bessel coefficient $J_n(x)$ is the coefficient of t^n in this expansion. If n is zero or a positive integer, we find that

$$J_n(x)=\sum_{s=0}^{\infty}\frac{(-1)^s}{s!(n+s)!}\left(\frac{x}{2}\right)^{n+2s}, \tag{27.2}$$

and when n is a negative integer we can deduce the series for $J_n(x)$ from equation (25.7).

Writing equation (27.2) in the form

$$J_n(x)=\frac{x^n}{2^n n!}\sum_{s=0}^{\infty}\frac{1}{s!(n+1)_s}(-\tfrac{1}{4}x^2)^s,$$

we see that

$$J_n(x)=\frac{x^n}{2^n n!}\,{}_0F_1(n+1;-\tfrac{1}{4}x^2). \tag{27.3}$$

The variation of the Bessel coefficients $J_0(x)$, $J_1(x)$, $J_2(x)$ for $0\leqslant x\leqslant 20$ is shown graphically in Fig. 9. These are the Bessel coefficients which occur most frequently in physical problems and their behaviour is similar to that of the general coefficient $J_n(x)$.

Simple relations for the Bessel coefficients may be derived easily from the series expansion (27.2). For example, since this equation is equivalent to

$$x^n J_n(x)=\sum_{s=0}^{\infty}\frac{(-1)^s x^{2n+2s}}{s!(n+s)!}\left(\frac{1}{2}\right)^{n+2s} \tag{27.4}$$

it follows, as a result of differentiating both sides of this equation with respect to x, and making use of the fact that $(2n+2s)/(n+s)!=2/(n-1+s)!$, that

$$\frac{\mathrm{d}}{\mathrm{d}x}\{x^n J_n(x)\}=\sum_{s=0}^{\infty}\frac{(-1)^s x^{2n+2s-1}}{s!(n-1+s)!}\left(\frac{1}{2}\right)^{n-1+2s}$$

which, by comparison with (27.4), shows that

$$\frac{\mathrm{d}}{\mathrm{d}x}\{x^n J_n(x)\}=x^n J_{n-1}(x). \tag{27.5}$$

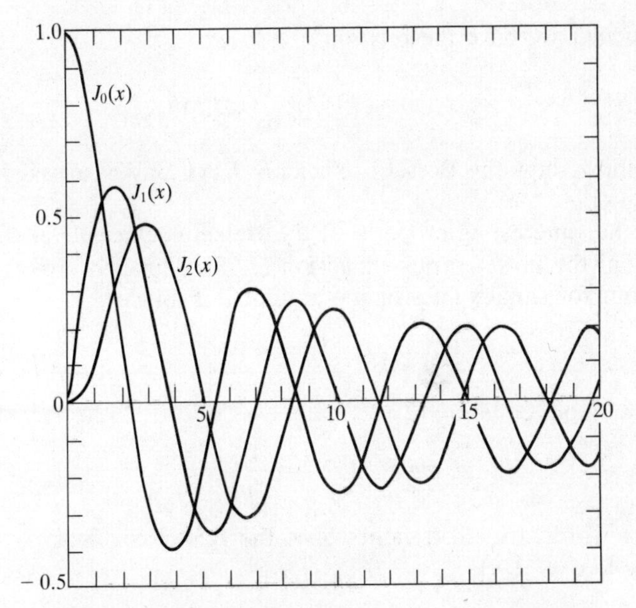

Fig. 9 Variation of $J_0(x)$, $J_1(x)$ and $J_2(x)$ with x

If we write this result in the form

$$\frac{1}{x}\frac{\mathrm{d}}{\mathrm{d}x}\{x^n J_n(x)\} = x^{n-1} J_{n-1}(x),$$

we see that if m is a positive integer less than n, then

$$\left(\frac{1}{x}\frac{\mathrm{d}}{\mathrm{d}x}\right)^m x^n J_n(x) = x^{n-m} J_{n-m}(x). \tag{27.6}$$

Similarly we can establish that

$$\frac{\mathrm{d}}{\mathrm{d}x}\{x^{-n} J_n(x)\} = -x^{-n} J_{n+1}(x) \tag{27.7}$$

or, which is the same thing,

$$\left(\frac{1}{x}\frac{\mathrm{d}}{\mathrm{d}x}\right)\{x^{-n} J_n(x)\} = -x^{-n-1} J_{n+1}(x),$$

a result which may be generalised to the form

$$\left(\frac{1}{x}\frac{\mathrm{d}}{\mathrm{d}x}\right)^m \{x^{-n} J_n(x)\} = (-1)^m x^{-n-m} J_{n+m}(x).$$

In particular we have the relation

$$x^{-n}J_n(x)=(-1)^n\left(\frac{1}{x}\frac{\mathrm{d}}{\mathrm{d}x}\right)^n J_0(x), \tag{27.8}$$

which shows how the Bessel coefficients $J_n(x)$ may be derived from $J_0(x)$.

Another interesting property of the Bessel coefficients also follows from the power series expansion (27.3). This concerns their behaviour for small values of the argument x. Since

$$\lim_{x\to 0} {}_0F_1(n+1, -\tfrac{1}{4}x^2)=1$$

it follows from equation (27.3) that

$$\lim_{x\to 0} x^{-n}J_n(x)=\frac{1}{2^n n!}. \tag{27.9}$$

In other words, for small values of x, the Bessel coefficient $J_n(x)$ behaves like $x^n/2^n n!$.

§28. Integral expressions for the Bessel coefficients

We have already derived one integral expression for the Bessel coefficient of order n (equation (25.4) above). In this section we shall consider other simple integral expressions for these coefficients.

We shall consider the integral

$$I=\int_{-1}^{1} (1-t^2)^{n-\frac{1}{2}}e^{ixt}\,\mathrm{d}t,$$

in which $n>-\frac{1}{2}$. If we develop $\exp(ixt)$ in ascending powers of ixt, we see that the value of this integral is

$$\sum_{s=0}^{\infty}\frac{(ix)^s}{s!}\int_{-1}^{1} (1-t^2)^{n-\frac{1}{2}}t^s\,\mathrm{d}t.$$

If s is an odd integer, then the corresponding integral occurring in this series is zero, and if s is an even integer, $2r$ say, then the integral has the value

$$\int_0^1 (1-u)^{n-\frac{1}{2}}u^{r-\frac{1}{2}}\,\mathrm{d}u=\frac{\Gamma(n+\frac{1}{2})\Gamma(r+\frac{1}{2})}{\Gamma(n+r+1)},$$

so that

$$I=\sum_{r=0}^{\infty}\frac{(-1)^r x^{2r}}{(2r)!}\frac{\Gamma(n+\frac{1}{2})\Gamma(r+\frac{1}{2})}{\Gamma(n+r+1)}$$

$$=\Gamma(\tfrac{1}{2})\Gamma(n+\tfrac{1}{2})\sum_{r=0}^{\infty}\frac{(-1)^r x^{2r}}{r!\Gamma(n+r+1)2^{2r}}$$

since, by the duplication formula for the gamma function,

$$\Gamma(\tfrac{1}{2})(2r)!=2^{2r}r!\Gamma(r+\tfrac{1}{2}).$$

It follows immediately from the series expansion (27.2) for $J_n(x)$ that

$$J_n(x)=\frac{(\frac{1}{2}x)^n}{\Gamma(\frac{1}{2})\Gamma(n+\frac{1}{2})}\int_{-1}^{1}(1-t^2)^{n-\frac{1}{2}}e^{ixt}\,dt, \tag{28.1}$$

and it is easily shown that this is equivalent to the formula

$$J_n(x)=\frac{x^n}{2^{n-1}\Gamma(\frac{1}{2})\Gamma(n+\frac{1}{2})}\int_{0}^{1}(1-t^2)^{n-\frac{1}{2}}\cos(xt)\,dt. \tag{28.2}$$

In particular

$$J_0(x)=\frac{2}{\pi}\int_0^1\frac{\cos(xt)}{\sqrt{(1-t^2)}}\,dt. \tag{28.3}$$

The result (28.2) may be expressed in a slightly different form by means of a simple change of variable. If we put $t=\cos\theta$, we obtain the integral expression

$$J_n(x)=\frac{x^n}{2^{n-1}\Gamma(\frac{1}{2})\Gamma(n+\frac{1}{2})}\int_0^{\frac{1}{2}\pi}\cos(x\cos\theta)\sin^{2n}\theta\,d\theta, \tag{28.4}$$

while if we make the substitution $t=\sin\theta$ we get the formula

$$J_n(x)=\frac{x^n}{2^{n-1}\Gamma(\frac{1}{2})\Gamma(n+\frac{1}{2})}\int_0^{\frac{1}{2}\pi}\cos(x\sin\theta)\cos^{2n}\theta\,d\theta. \tag{28.5}$$

The particular forms appropriate to $n=0$ are

$$J_0(x)=\frac{2}{\pi}\int_0^{\frac{1}{2}\pi}\cos(x\cos\theta)\,d\theta=\frac{2}{\pi}\int_0^{\frac{1}{2}\pi}\cos(x\sin\theta)\,d\theta. \tag{28.6}$$

§29. The addition formula for the Bessel coefficients

In certain physical problems we have to reduce a Bessel coefficient of type $J_n(x+y)$ to a form more amenable to computation.

We shall now derive an addition formula which is of great use in these circumstances. From the definition (25.6) we have the expansion

$$\exp\left\{\tfrac{1}{2}(x+y)\left(t-\frac{1}{t}\right)\right\}=\sum_{n=-\infty}^{\infty} J_n(x+y)t^n.$$

Writing the left-hand side as a product

$$\exp\left\{\tfrac{1}{2}x\left(t-\frac{1}{t}\right)\right\}\exp\left\{\tfrac{1}{2}y\left(t-\frac{1}{t}\right)\right\}$$

and inserting the appropriate series from (25.6), we find that

$$\sum_{n=-\infty}^{\infty} J_n(x+y)t^n=\sum_{r=-\infty}^{\infty}\sum_{s=-\infty}^{\infty} J_r(x)J_s(y)t^{r+s}.$$

Equating coefficients of t^n we obtain the addition formula

$$J_n(x+y)=\sum_{r=-\infty}^{\infty} J_r(x)J_{n-r}(y). \tag{29.1}$$

To put this in a form which involves only Bessel coefficients of positive order, we write the right-hand side in the form

$$\sum_{r=-\infty}^{-1} J_r(x)J_{n-r}(y)+\sum_{r=0}^{n} J_r(x)J_{n-r}(y)+\sum_{r=n+1}^{\infty} J_r(x)J_{n-r}(y)$$

and note that because of the relation (25.7) the first term can be written as

$$\sum_{r=-\infty}^{-1}(-1)^r J_{-r}(x)J_{n-r}(y)\equiv\sum_{r=1}^{\infty}(-1)^r J_r(x)J_{n+r}(y).$$

Similarly the third term is equal to

$$\sum_{r=1}^{\infty} J_{n+r}(x)J_{-r}(y)=\sum_{r=1}^{\infty}(-1)^r J_{n+r}(x)J_r(y),$$

so that finally we have

$$J_n(x+y)=\sum_{r=0}^{n} J_r(x)J_{n-r}(y)$$
$$+\sum_{r=1}^{\infty}(-1)^r\{J_r(x)J_{n+r}(y)+J_{n+r}(x)J_r(y)\}. \tag{29.2}$$

§30. Bessel's differential equation

We showed previously (§26 above) that, if n is an integer, $J_n(x)$ is a solution of Bessel's equation (26.9). We shall now examine the solutions of that equation when the parameter n is not necessarily an integer. To emphasise that this parameter is, in general, non-integral, we shall replace it by the symbol ν, so that we now consider the solutions of the second order linear differential equation

$$\frac{d^2y}{dx^2}+\frac{1}{x}\frac{dy}{dx}+\left(1-\frac{\nu^2}{x^2}\right)y=0. \tag{30.1}$$

Writing the equation in the form

$$x^2\frac{d^2y}{dx^2}+x\frac{dy}{dx}+(-\nu^2+x^2)y=0, \tag{30.2}$$

we see that the point $x=0$ is a regular singular point and that in the notation of §3, $p_0=1$ and $q_0=-\nu^2$. The indicial equation (cf. (2.10) above) is therefore

$$\rho^2-\nu^2=0$$

and this has roots $\rho=\pm\nu$.

First of all we shall suppose that ν is neither zero nor an integer. Then the first solution is of the form

$$y=\sum_{r=0}^{\infty} c_r x^{r+\nu}. \tag{30.3}$$

Substituting this series in equation (30.2) we see that the coefficients c_r must be such that

$$\sum_{r=0}^{\infty}\{(\nu+r)(\nu+r-1)+(\nu+r)-\nu^2\}c_r x^{r+\nu}+\sum_{r=0}^{\infty} c_r x^{r+\nu+2}=0.$$

Hence we must have

$$c_1\{(\nu+1)^2-\nu^2\}=0$$

and in general

$$c_r\{(r+\nu)^2-\nu^2\}=-c_{r-2}, \qquad r=2,3,\ldots. \tag{30.4}$$

We must therefore take c_1 to be zero and hence, in order that (30.4) may be satisfied for all $r\geqslant 2$, we must take

$$c_{2r+1}=0$$

and

$$c_{2r}=\frac{(-1)^r c_0}{(2r+2\nu)(2r+2\nu-2)\ldots(2\nu+2)2r(2r-2)\ldots 2},$$

an expression which may be put in the form

$$c_{2r}=\frac{c_0}{r!(\nu+1)_r}\left(-\frac{1}{4}\right)^r.$$

Taking $c_0=1/2^\nu\nu!$ we see that the basic solution of type (30.3) may be taken as

$$y=\frac{x^\nu}{2^\nu\Gamma(\nu+1)}\sum_{r=0}^{\infty}\frac{(-\frac{1}{4}x^2)^r}{r!(\nu+1)_r}. \tag{30.5}$$

Comparing this series with series (27.2) we see that it is of precisely the same form as that equation, the only difference being that n is replaced here by ν. If we take the series (27.2) to define the **Bessel function of the first kind of order *n*,** even when n is not an integer, then we may write the solution (30.5) in the form

$$y=J_\nu(x).$$

Similarly, if we substitute a series of type

$$y=\sum_{r=0}^{\infty}c_r x^{r-\nu}$$

to correspond to the second root of the indicial equation, we find that it must be of the type

$$y=\frac{x^{-\nu}}{2^\nu\Gamma(-\nu+1)}\sum_{r=0}^{\infty}\frac{(-\frac{1}{4}x^2)^r}{r!(-\nu+1)_r} \tag{30.6}$$

and with the extension of the definition (27.2) to non-integral values of ν we may write this solution in the form

$$y=J_{-\nu}(x).$$

Thus when ν is *not* an integer we may write the general solution of equation (30.1) in the form

$$y=AJ_\nu(x)+BJ_{-\nu}(x), \tag{30.7}$$

where $J_\nu(x)$ is defined by the equation

$$J_\nu(x)=\frac{x^\nu}{2^\nu\Gamma(\nu+1)}\,{}_0F_1(\nu+1;-\tfrac{1}{4}x^2). \tag{30.8}$$

It should be observed that the results of §§27, 28, with the exception of (27.8), are true when n is not an integer, since they were derived directly from the definition (27.2), which is equivalent to (30.8). The transition is effected merely by replacing $n!$ by $\Gamma(\nu+1)$.

When ν is zero or an integer we know from equation (25.7) that the solutions $J_\nu(x)$ and $J_{-\nu}(x)$ are not linearly independent. We must therefore use the formulae (3.8) to calculate the second solution.

We shall consider first the case in which $\nu=0$. If we let

$$w=\sum_{r=0}^{\infty} c_r x^{r+\rho},$$

then in order to satisfy the recurrence relation (30.4) we must have

$$w=x^\rho \sum_{r=0}^{\infty} \frac{(-\frac{1}{4}x)^r}{r!(\rho+1)_r} \tag{30.9}$$

and putting $\rho=0$ we obtain the first solution

$$w_0=J_0(x). \tag{30.10}$$

Using the result

$$\frac{\partial}{\partial\rho}\frac{1}{(\rho+1)_r}=-\frac{1}{(\rho+1)_r}\left\{\sum_{s=1}^{r}\frac{1}{\rho+s}\right\}$$

we see that

$$\frac{\partial w}{\partial\rho}=w\log x-x^\rho\sum_{r=1}^{\infty}\frac{(-\frac{1}{4}x)^r}{r!(\rho+1)_r}\left\{\sum_{s=1}^{r}\frac{1}{\rho+s}\right\}.$$

Putting $\rho=0$ and substituting the value (30.10) for w_0 we find that the second solution $(\partial w/\partial\rho)_{\rho=0}$ is

$$\mathsf{Y}_0(x)=J_0(x)\log x-\sum_{r=1}^{\infty}\frac{(-\frac{1}{4}x)^r}{(r!)^2}\phi(r), \tag{30.11}$$

where

$$\phi(r)=\sum_{s=1}^{r}\frac{1}{s}. \tag{30.12}$$

The function $\mathsf{Y}_0(x)$ so obtained is called **Neumann's Bessel function of the second kind of zero order.** Obviously if we add to

$\mathrm{Y}_0(x)$ a function which is a constant multiple of $J_0(x)$ the resulting function is also a solution of the differential equation

$$\frac{d^2y}{dx^2}+\frac{1}{x}\frac{dy}{dx}+y=0. \tag{30.13}$$

In particular the function

$$Y_0(x)=\frac{2}{\pi}\{\mathrm{Y}_0(x)-(\log 2-\gamma)J_0(x)\},$$

where γ denotes Euler's constant, is a second solution of the equation. Substituting from equation (30.11) for $\mathrm{Y}_0(x)$ in this equation we obtain the expression

$$Y_0(x)=\frac{2}{\pi}\{\log(\tfrac{1}{2}x)+\gamma\}J_0(x)-\frac{2}{\pi}\sum_{r=1}^{\infty}\frac{(-\frac{1}{4}x^2)^r}{(r!)^2}\phi(r), \tag{30.14}$$

where $\phi(r)$ is defined by equation (30.12).

The function $Y_0(x)$ so defined is known as **Weber's Bessel function of the second kind of zero order.**

Thus the complete solution of the equation (30.13) is

$$y=AJ_0(x)+BY_0(x), \tag{30.15}$$

where A, B are arbitrary constants and $J_0(x)$, $Y_0(x)$ are given by equations (30.8) and (30.14) respectively.

It can be shown by an exactly similar process that when ν is an integer the complete solution of the equation (30.1) is

$$y=AJ_\nu(x)+BY_\nu(x), \tag{30.16}$$

where A, B are arbitrary constants, $J_\nu(x)$ is defined by equation (30.8) and $Y_\nu(x)$ is given by

$$Y_\nu(x)=\frac{2}{\pi}\{\gamma+\log(\tfrac{1}{2}x)\}J_\nu(x)-\frac{1}{\pi}\sum_{r=0}^{\nu-1}\frac{(\nu-r-1)!}{r!}\left(\frac{2}{x}\right)^{\nu-2r}$$

$$-\frac{1}{\pi}\sum_{r=0}^{\infty}\frac{(-1)^r(\frac{1}{2}x)^{\nu+2r}}{r!(\nu+r)!}\{\phi(r+\nu)+\phi(r)\}. \tag{30.17}$$

The function $Y_\nu(x)$ so defined† reduces to the $Y_0(x)$ of equation (30.14) as $\nu\to 0$ and is known as **Weber's Bessel function of the second kind of order ν.**

The variation of $Y_0(x)$ and $Y_1(x)$ for a range of values of x is shown graphically in Fig. 10.

†This function is denoted as $N_\nu(x)$ by Courant and Hilbert.

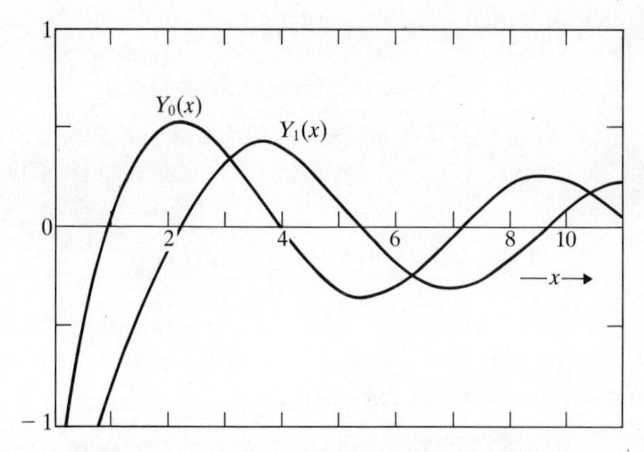

Fig. 10 Variation of $Y_0(x)$ and $Y_1(x)$ with x

The functions $J_\nu(x)$ and $Y_\nu(x)$ are independent solutions of the equation (30.1), but in certain circumstances it is advantageous to define, in terms of them, two new independent solutions. If we write

$$H_\nu^{(1)}(x) = J_\nu(x) + iY_\nu(x), \tag{30.18}$$

$$H_\nu^{(2)}(x) = J_\nu(x) - iY_\nu(x), \tag{30.19}$$

then it is obvious that we can take the general solution of Bessel's differential equation (30.1) to be

$$y = A_1 H_\nu^{(1)}(x) + A_2 H_\nu^{(2)}(x), \tag{30.20}$$

where A_1 and A_2 are arbitrary constants. The functions $H_\nu^{(1)}(x)$, $H_\nu^{(2)}(x)$ defined by equations (30.18) and (30.19) are called **Hankel's Bessel functions of the third kind of order ν.**

The Hankel functions $H_\nu^{(1)}(x)$ and $H_\nu^{(2)}(x)$ bear the same relation to the Bessel functions $J_\nu(x)$, $Y_\nu(x)$ as the functions $\exp(\pm i\nu x)$ bear to $\cos \nu x$ and $\sin \nu x$, and are used in analysis for similar reasons. It should also be noted that the Bessel functions $Y_\nu(x)$, $H_\nu^{(1)}(x)$, $H_\nu^{(2)}(x)$ satisfy the same differential equations and recurrence relations as the function $J_\nu(x)$.

§31. Spherical Bessel functions

A problem which arises in mathematical physics is that of the solution of the wave equation in spherical polar coordinates

$$\frac{\partial^2\psi}{\partial r^2} + \frac{2}{r}\frac{\partial\psi}{\partial r} + \frac{1}{r^2 \sin\theta}\frac{\partial}{\partial\theta}\left(\sin\theta\frac{\partial\psi}{\partial\theta}\right) + \frac{1}{r^2\sin^2\theta}\frac{\partial^2\psi}{\partial\phi^2} = \frac{1}{c^2}\frac{\partial^2\psi}{\partial t^2}. \tag{31.1}$$

If we take a solution of this equation of the form

$$\psi = Y_{m,n}(\theta, \phi)\psi(r)e^{i\omega t}, \tag{31.2}$$

where $Y_{m,n}(\theta, \phi)$ is the surface spherical harmonic defined by equation (23.3) and $\psi(r)$ is a function of r alone which satisfies the equation

$$\frac{d^2\Psi}{dr^2}+\frac{2}{r}\frac{d\Psi}{dr}-\frac{n(n+1)}{r^2}\Psi+\frac{\omega^2}{c^2}\Psi=0. \tag{31.3}$$

Now putting

$$\Psi = r^{-\frac{1}{2}}R \tag{31.4}$$

we see that equation (31.3) becomes

$$\frac{d^2R}{dr^2}+\frac{1}{r}\frac{dR}{dr}+\left\{\frac{\omega^2}{c^2}-\frac{(n+\frac{1}{2})^2}{r^2}\right\}R=0,$$

whose general solution is readily seen to be

$$R = AJ_{n+\frac{1}{2}}(\omega r/c)+BJ_{-n-\frac{1}{2}}(\omega r/c). \tag{31.5}$$

Hence the function

$$\psi = r^{-\frac{1}{2}}Y_{m,n}(\theta, \phi)J_{\pm(n+\frac{1}{2})}(\omega r/c)e^{i\omega t} \tag{31.6}$$

is a solution of the equation (31.1).

The functions $J_{\pm(n+\frac{1}{2})}(x)$ which occur in the solution (31.6) are called **spherical Bessel functions.** We shall now show that they are related simply to the circular functions. First of all we consider the Bessel function $J_{\frac{1}{2}}(x)$. If we let $\nu=\frac{1}{2}$ in equation (30.6) and make use of the duplication formula for the gamma function we obtain the result

$$J_{\frac{1}{2}}(x)=\left(\frac{2}{\pi x}\right)^{\frac{1}{2}}\sum_{r=0}^{\infty}\frac{(-1)^r x^{2r+1}}{(2r+1)!},$$

which shows that

$$J_{\frac{1}{2}}(x)=\left(\frac{2}{\pi x}\right)^{\frac{1}{2}}\sin x. \tag{31.7}$$

Again, if we put $\nu=-\frac{1}{2}$ in equation (30.6) we obtain the relation

$$J_{-\frac{1}{2}}(x)=\left(\frac{2}{\pi x}\right)^{\frac{1}{2}}\cos x. \tag{31.8}$$

Table 1

m	f_m	g_m
$\frac{3}{2}$	$\frac{1}{x}$	1
$\frac{5}{2}$	$\frac{3}{x^2}-1$	$\frac{3}{x}$
$\frac{7}{2}$	$\frac{15}{x^3}-\frac{6}{x}$	$\frac{15}{x^2}-1$
$\frac{9}{2}$	$\frac{105}{x^4}-\frac{45}{x^2}+1$	$\frac{105}{x^3}-\frac{10}{x}$
$\frac{11}{2}$	$\frac{945}{x^5}-\frac{420}{x^3}+\frac{15}{x}$	$\frac{945}{x^4}-\frac{105}{x^2}+1$

The other functions $J_m(x)$, where m is half an odd integer, may be worked out in a similar fashion. It is left as an exercise to the reader to show that

$$J_m(x)=\left(\frac{2}{\pi x}\right)^{\frac{1}{2}}\{f_m(x)\sin x-g_m(x)\cos x\},$$

$$J_{-m}(x)=\left(\frac{2}{\pi x}\right)^{\frac{1}{2}}(-1)^{n-\frac{1}{2}}\{g_m(x)\sin x+f_m(x)\cos x\},$$

where the functions f_m, g_m are given in Table 1. These functions which arise in the way described have been tabulated in *Tables of Spherical Bessel Functions* (1947).

§32. Integrals involving Bessel functions

In this section we shall derive the values of some integrals involving Bessel functions which arise in practical applications. In the first instance we shall consider definite integrals.

From equation (27.5) we have the relation

$$\int_0^{\alpha} x^n J_{n-1}(x)\,dx=[x^n J_n(x)]_0^{\alpha}.$$

If $n>0$, $x^n J_n(x)\to 0$ as $x\to 0$, so that the lower limit is zero and we obtain the integral

$$\int_0^{\alpha} x^n J_{n-1}(x)\,dx=\alpha^n J_n(\alpha),\qquad n>0,\tag{32.1}$$

which, by a simple change of variable, gives the result

$$\int_0^a r^n J_{n-1}(\xi r)\,dr = \frac{a^n}{\xi} J_n(\xi a), \qquad n > 0. \tag{32.2}$$

A particular case of this result which is of frequent use in mathematical physics is obtained by putting $n = 1$ in equation (32.2). In this way we obtain the integral

$$\int_0^a r J_0(\xi r)\,dr = \frac{a}{\xi} J_1(a\xi). \tag{32.3}$$

Further results may be obtained from (32.2) by familiar devices such as integration by parts. For example, making use of equation (27.5) we may write

$$\int_0^a r^3 J_0(\xi r)\,dr = \int_0^a r^2 \frac{1}{\xi}\frac{\partial}{\partial r}\{rJ_1(\xi r)\}\,dr,$$

and, integrating by parts, we see that the right-hand side of this equation becomes

$$\frac{a^3}{\xi} J_1(\xi a) - \frac{2}{\xi}\int_0^a r^2 J_1(\xi r)\,dr$$

which reduces, by virtue of (32.2), to

$$\frac{a^3}{\xi} J_1(\xi a) - \frac{2a^2}{\xi^2} J_2(\xi a).$$

Now by the recurrence relation (26.2) we have the expression

$$J_2(\xi a) = \frac{2}{\xi a} J_1(\xi a) - J_0(\xi a),$$

so that finally we have the result

$$\int_0^a r^3 J_0(\xi r)\,dr = \frac{2a^2}{\xi^2}\left\{J_0(\xi a) + \left(\tfrac{1}{2}a\xi - \frac{2}{a\xi}\right) J_1(a\xi)\right\}. \tag{32.4}$$

Combining this result with equation (32.3) we obtain the integral

$$\int_0^a r(a^2 - r^2) J_0(\xi r)\,dr = \frac{4a}{\xi^3} J_1(\xi a) - \frac{2a^2}{\xi^2} J_0(\xi a). \tag{32.5}$$

The most commonly occurring infinite integrals are most easily evaluated by means of substituting the formula (27.3) in parts (ii) and (iii) of Problem 2.17. From part (ii) we see that

$$\frac{a^\nu}{2^\nu\Gamma(\nu+1)}\int_0^\infty {}_0F_1(\nu+1;-\tfrac{1}{4}a^2x^2)x^{\nu+\mu}e^{-px}\,dx$$

$$=\frac{\Gamma(\mu+\nu+1)a^\nu}{2^\nu\Gamma(\nu+1)p^{\mu+\nu+1}}\,{}_2F_1\left(\tfrac{1}{2}\mu+\tfrac{1}{2}\nu+\tfrac{1}{2},\tfrac{1}{2}\mu+\tfrac{1}{2}\nu+1;\nu+1;-\frac{a^2}{p^2}\right). \quad (32.6)$$

If we make use of equation (30.8) on the left-hand side of this equation and of equation (7.4) on the right-hand side, we see that this result is equivalent to the formula

$$\int_0^\infty J_\nu(ax)x^\mu e^{-px}\,dx=\frac{\Gamma(\mu+\nu+1)a^\nu}{2^\nu\Gamma(\nu+1)(a^2+p^2)^{\frac{1}{2}\mu+\frac{1}{2}\nu+\frac{1}{2}}}$$

$$\times{}_2F_1\left(\tfrac{1}{2}\mu+\tfrac{1}{2}\nu+\tfrac{1}{2};\tfrac{1}{2}\nu-\tfrac{1}{2}\mu;\nu+1;\frac{a^2}{a^2+p^2}\right), \quad (32.7)$$

where $p>0$, $\mu+\nu>0$.

The hypergeometric series occurring on the right-hand side of this equation assumes a particularly simple form if either $\mu=\nu$ or $\mu=\nu+1$, and we obtain the formulae

$$\int_0^\infty J_\nu(ax)x^\nu e^{-px}\,dx=\frac{2^\nu\Gamma(\nu+\frac{1}{2})}{\Gamma(\frac{1}{2})}\frac{a^\nu}{(a^2+p^2)^{\nu+\frac{1}{2}}}, \quad (32.8)$$

$$\int_0^\infty J_\nu(ax)x^{\nu+1}e^{-px}\,dx=\frac{2^{\nu+1}\Gamma(\nu+\frac{3}{2})}{\Gamma(\frac{1}{2})}\frac{pa^\nu}{(a^2+p^2)^{\nu+\frac{3}{2}}}. \quad (32.9)$$

Two special cases of the formula (32.8) which occur frequently are

$$\int_0^\infty J_0(ax)e^{-px}\,dx=\frac{1}{\sqrt{(a^2+p^2)}}, \quad (32.10)$$

$$\int_0^\infty xJ_1(ax)e^{-px}\,dx=\frac{a}{(a^2+p^2)^{\frac{3}{2}}}. \quad (32.11)$$

Integrating both sides of equation (32.11) with respect to p from p to ∞ we find that

$$\int_0^\infty J_1(ax)e^{-px}\,dx=\frac{1}{a}-\frac{p}{a\sqrt{(a^2+p^2)}}. \quad (32.12)$$

A special case of (32.9) which is often needed is

$$\int_0^\infty xJ_0(ax)e^{-px}\,dx=\frac{p}{(a^2+p^2)^{\frac{3}{2}}}. \quad (32.13)$$

If we let p tend to zero on both sides of equation (32.7) we find that we can sum the hypergeometric series by Gauss's theorem (7.2) provided that $|\mu|<|\nu+1|$. We then have the result

$$\int_0^\infty J_\nu(ax)x^\mu\,dx=\frac{2^\mu\Gamma(\frac{1}{2}+\frac{1}{2}\mu+\frac{1}{2}\nu)}{a^{\mu+1}\Gamma(\frac{1}{2}-\frac{1}{2}\mu+\frac{1}{2}\nu)}. \tag{32.14}$$

Similarly from Problem 2.17(iii) we have the equation

$$\int_0^\infty {}_0F_1(\nu+1;-\tfrac{1}{4}a^2x^2)e^{p^2x^2}x^{\mu+\nu-1}\,dx$$
$$=\frac{\Gamma(\frac{1}{2}\mu+\frac{1}{2}\nu)}{2p^{\mu+\nu}}\,{}_1F_1\left(\tfrac{1}{2}\mu+\tfrac{1}{2}\nu;\nu+1;-\frac{a^2}{4p^2}\right)$$

which, because of (27.3), is equivalent to

$$\int_0^\infty J_\nu(ax)e^{-p^2x^2}x^{\mu-1}\,dx$$
$$=\frac{a^\nu\Gamma(\frac{1}{2}\mu+\frac{1}{2}\nu)}{2^{\nu+1}p^{\mu+\nu}\Gamma(\nu+1)}\,{}_1F_1\left(\tfrac{1}{2}\mu+\tfrac{1}{2}\nu;\nu+1;-\frac{a^2}{4p^2}\right). \tag{32.15}$$

From Problem 2.11(i) and (ii) we deduce the special cases

$$\int_0^\infty x^{\nu+1}J_\nu(ax)e^{-p^2x^2}\,dx=\frac{a^\nu e^{-a^2/4p^2}}{(2p^2)^{\nu+1}}, \tag{32.16}$$

and

$$\int_0^\infty x^{\nu+3}J_\nu(ax)e^{-p^2x^2}\,dx=\frac{a^\nu}{2^{\nu+1}p^{2\nu+4}}\left(\nu+1-\frac{a^2}{4p^2}\right)e^{-a^2/4p^2} \tag{32.17}$$

of which the most frequently used are

$$\int_0^\infty xJ_0(ax)e^{-p^2x^2}\,dx=\frac{1}{2p^2}\,e^{-a^2/4p^2} \tag{32.18}$$

and

$$\int_0^\infty x^3J_0(ax)e^{-p^2x^2}\,dx=\frac{1}{2p^4}\left(1-\frac{a^2}{4p^2}\right)e^{-a^2/4p^2}. \tag{32.19}$$

A result from the theory of integral transforms is most useful in the evaluation of integrals involving Bessel functions of the first kind. It is called the Hankel inversion theorem and we shall state it without proof:

If

$$\int_0^\infty xf(x)J_\nu(xt)\,dx=g(t),\qquad \nu>-\tfrac{1}{2},$$

then

$$\int_0^\infty tg(t)J_\nu(xt)\,dt = f(x).$$

For example in (32.2), if we replace ξ by t and r by x we obtain

$$\int_0^\infty x\{x^\nu H(a-x)\}J_\nu(xt)\,dx = a^{-\nu-1}t^{-1}J_{\nu+1}(ta),$$

where H is the Heaviside unit function defined by the relations

$$H(z) = \begin{cases} 1 & \text{if } z>0, \\ 0 & \text{if } z<0. \end{cases}$$

Applying the Hankel inversion theorem we deduce immediately that

$$\int_0^\infty J_{\nu+1}(ta)J_\nu(tx)\,dt = x^\nu a^{-\nu-1}H(a-x). \tag{32.20}$$

In a similar way we can deduce from equation (32.8) that

$$\int_0^\infty t^{\nu+1}(t^2+p^2)^{-\nu-\frac{1}{2}}J_\nu(xt)\,dt = \frac{\Gamma(\frac{1}{2})}{2^\nu\Gamma(\nu+\frac{1}{2})}x^{\nu-1}e^{-px}. \tag{32.21}$$

§33. The modified Bessel functions

By an argument similar to that employed in §1 we can readily show that Laplace's equation in cylindrical coordinates

$$\frac{\partial^2\psi}{\partial\rho^2}+\frac{1}{\rho}\frac{\partial\psi}{\partial\rho}+\frac{1}{\rho^2}\frac{\partial^2\psi}{\partial\phi^2}+\frac{\partial^2\psi}{\partial z^2}=0$$

possesses solutions of the form

$$\psi = e^{\pm i\nu\phi\pm imz}R(\rho),$$

where $R(\rho)$ satisfies the ordinary differential equation

$$\frac{d^2R}{d\rho^2}+\frac{1}{\rho}\frac{dR}{d\rho}-\left(m^2+\frac{\nu^2}{\rho^2}\right)R=0. \tag{33.1}$$

Writing x in place of $m\rho$ we see that this equation is equivalent to the equation

$$\frac{d^2R}{dx^2}+\frac{1}{x}\frac{dR}{dx}-\left(1+\frac{\nu^2}{x^2}\right)R=0. \tag{33.2}$$

If we proceed in exactly the same way as in §30 we can show that if ν is neither zero nor an integer the solution of this equation is

$$R = AI_\nu(x) + BI_{-\nu}(x), \tag{33.3}$$

where A and B are arbitrary constants and the function $I_\nu(x)$ is defined by the equation

$$\begin{aligned} I_\nu(x) &= \frac{x^\nu}{2^\nu\Gamma(\nu+1)} \sum_{r=0}^{\infty} \frac{(\frac{1}{4}x^2)^r}{r!(\nu+1)_r} \\ &= \frac{x^\nu}{2^\nu\Gamma(\nu+1)}\, {}_0F_1(\nu+1; \tfrac{1}{4}x^2). \end{aligned} \tag{33.4}$$

Comparing equation (33.4) with equation (30.8) we see that

$$I_\nu(x) = i^{-\nu}J_\nu(ix) \tag{33.5}$$

a result which might have been conjectured from the differential equation itself.

If ν is an integer, n say, then $I_{-n}(x)$ is a multiple of $I_n(x)$, so that the solution (33.3) in effect contains only one arbitrary constant. By a process similar to that outlined in §30 we can show that in these circumstances the general solution of equation (34.2) is

$$R = AI_n(x) + BK_n(x), \tag{33.6}$$

where the function $K_n(x)$ is defined by the equation

$$\begin{aligned} K_n(x) &= (-1)^{n+1}I_n(x)\{\log(\tfrac{1}{2}x) + \gamma\} \\ &\quad + \tfrac{1}{2}\sum_{r=0}^{n-1} \frac{(-1)^r(n-r-1)!}{r!}(\tfrac{1}{2}x)^{-n+2r} \\ &\quad + \tfrac{1}{2}(-1)^n \sum_{r=1}^{\infty} \frac{1}{r!(n+r)!}\{\phi(r) + \phi(n+r)\}(\tfrac{1}{2}x)^{n+2r}. \end{aligned} \tag{33.7}$$

The functions $I_n(x)$, $K_n(x)$ defined by equations (34.4) and (34.7) respectively are known as **modified Bessel functions** of the first and second kinds.

The result (33.5) is very useful for deducing properties of the modified Bessel function $I_n(x)$ from those of the Bessel function $J_n(x)$. For instance, when n is an integer it follows from equation (25.7) that

$$I_{-n}(x) = I_n(x), \tag{33.8}$$

and from equations (26.1) to (26.5) respectively that

$$2I'_n(x)=I_{n-1}(x)+I_{n+1}(x), \tag{33.9}$$

$$\frac{2n}{x}I_n(x)=I_{n-1}(x)-I_{n+1}(x), \tag{33.10}$$

$$xI'_n(x)=xI_{n-1}(x)-nI_n(x), \tag{33.11}$$

$$xI'_n(x)=nI_n(x)+xI_{n+1}(x), \tag{33.12}$$

$$I'_0(x)=I_1(x). \tag{33.13}$$

Similarly equations (27.5) and (27.7) imply the relations

$$\frac{\mathrm{d}}{\mathrm{d}x}\{x^nI_n(x)\}=x^nI_{n-1}(x), \tag{33.14}$$

$$\frac{\mathrm{d}}{\mathrm{d}x}\{x^{-n}I_n(x)\}=x^{-n}I_{n+1}(x). \tag{33.15}$$

All of these relations can, of course, be derived directly from the definition (33.4) of $I_\nu(x)$, and it is suggested as an exercise to the reader to derive them in this way.

It should also be observed that $K_n(x)$ satisfies the same recurrence relations as $I_n(x)$.

The variation of $I_0(x)$, $I_1(x)$ and $I_2(x)$ with x is shown graphically in Fig. 11 and that of $K_0(x)$, $K_1(x)$ and $K_2(x)$ is shown in Fig. 12.

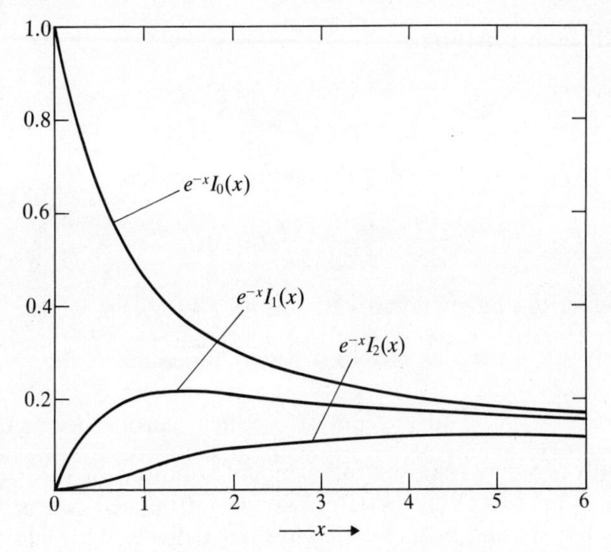

Fig. 11 Variation of $e^{-x}I_0(x)$, $e^{-x}I_1(x)$ and $e^{-x}I_2(x)$ with x

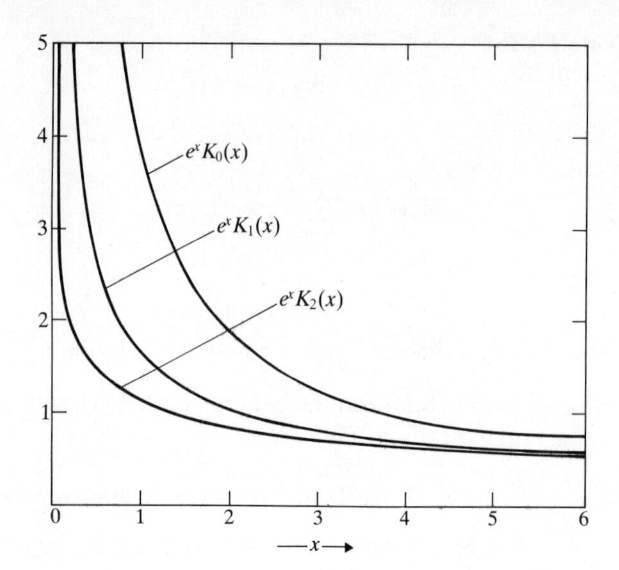

Fig. 12 Variation of $e^x K_0(x)$, $e^x K_1(x)$ and $e^x K_2(x)$ with x

§34. The Ber and Bei functions

If we wish to find solutions of the form

$$\psi = R(\rho)e^{i\omega t}$$

of the diffusion equation

$$\frac{\partial^2 \psi}{\partial \rho^2} + \frac{1}{\rho}\frac{\partial \psi}{\partial \rho} = \frac{1}{\kappa}\frac{\partial \psi}{\partial t},$$

we have to solve the ordinary differential equation

$$\frac{d^2 R}{d\rho^2} + \frac{1}{\rho}\frac{dR}{d\rho} - \frac{i\omega}{\kappa}R = 0.$$

On changing the independent variable to $x = (\omega/\kappa)^{\frac{1}{2}}\rho$, we see that this latter equation is equivalent to the equation

$$\frac{d^2 R}{dx^2} + \frac{1}{x}\frac{dR}{dx} - iR = 0. \tag{34.1}$$

Formally we may take the independent solutions of this equation to be $I_0(i^{\frac{1}{2}}x)$ and $K_0(i^{\frac{1}{2}}x)$. Kelvin introduced two new functions ber (x) and bei (x) which are respectively the real and imaginary parts of $I_0(i^{\frac{1}{2}}x)$, i.e.

$$\text{ber}\,(x)+i\,\text{bei}\,(x)=I_0(i^{\frac{1}{2}}x). \tag{34.2}$$

From the definition (33.4) of $I_0(x)$ we see that

$$\text{ber}\,(x)=\sum_{s=0}^{\infty}\frac{(-1)^s(\frac{1}{4}x^2)^{2s}}{(2s!)^2}, \tag{34.3}$$

and that

$$\text{bei}\,(x)=\sum_{s=0}^{\infty}\frac{(-1)^s(\frac{1}{4}x^2)^{2s+1}}{(2s+1)!^2}. \tag{34.4}$$

The variation of the functions ber (x) and bei (x) with x is shown diagrammatically in Fig. 13.

In a similar way the functions ker (x) and kei (x) are defined to be respectively the real and imaginary parts of the complex function $K_0(i^{\frac{1}{2}}x)$, i.e.

$$\text{ker}\,(x)+i\,\text{kei}\,(x)=K_0(i^{\frac{1}{2}}x). \tag{34.5}$$

From the definition (33.7) of $K_0(x)$ we can readily show that

$$\text{ker}\,(x)=-\{\log(\tfrac{1}{2}x)+\gamma\}\,\text{ber}\,(x)+\tfrac{1}{4}\pi\,\text{bei}\,(x)$$

$$+\sum_{r=1}^{\infty}\frac{(-1)^r(\frac{1}{2}x)^{4r}}{(2r)!^2}\,\phi(2r), \tag{34.6}$$

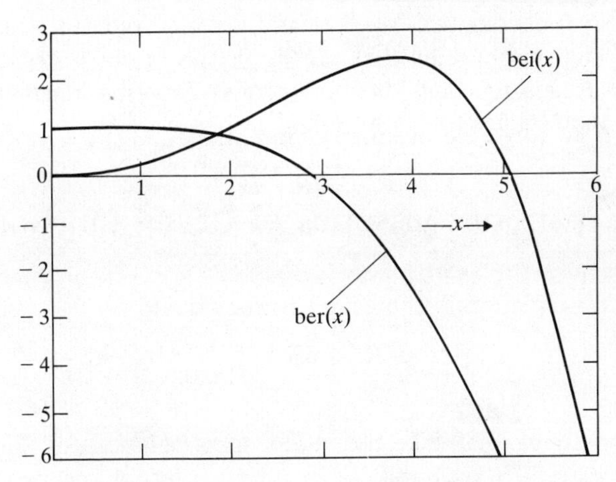

Fig. 13 Variation of ber (x) and bei (x) with x

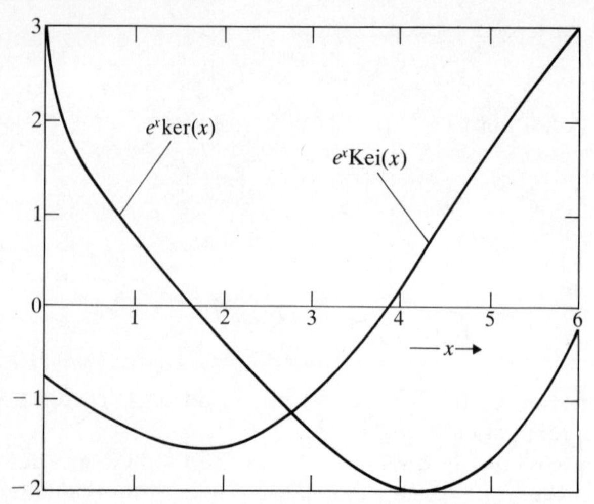

Fig. 14 Variation of ker (x) and kei (x) with x

and that

$$\operatorname{kei}(x) = -\{\log(\tfrac{1}{2}x)+\gamma\}\operatorname{bei}(x) - \tfrac{1}{4}\pi \operatorname{ber}(x) + \sum_{r=0}^{\infty} \frac{(-1)^r(\frac{1}{2}x)^{4r+2}}{(2r+1)!^2}\phi(2r+1). \tag{34.7}$$

Fig. 14 shows the variation of the functions ker (x) and kei (x) over a range of values of the independent variable x.

The four functions ber, bei, ker and kei are used more often in electrical engineering than they are in physics or chemistry. They are called *Kelvin functions* or (more rarely) *Thomson functions*; tables of their values are to be found in Nosova (1961).

§35. Expansions in series of Bessel functions

We know from §30 that

$$\left\{x^2\frac{d^2}{dx^2} + x\frac{d}{dx} + (\lambda^2x^2 - m^2)\right\}J_m(\lambda x) = 0, \tag{35.1}$$

$$\left\{x^2\frac{d^2}{dx^2} + x\frac{d}{dx} + (\mu^2x^2 - n^2)\right\}J_n(\mu x) = 0, \tag{35.2}$$

so that multiplying equation (35.1) by $J_n(\mu x)/x$, (35.2) by

$J_m(\lambda x)/x$, integrating with respect to x from 0 to a and subtracting we find that

$$(\lambda^2-\mu^2)\int_0^a xJ_m(\lambda x)J_n(\mu x)\,\mathrm{d}x+(n^2-m^2)\int_0^a J_m(\lambda x)J_n(\mu x)\frac{\mathrm{d}x}{x}$$
$$=a[\mu J_n(\lambda a)J'_m(\mu a)-\lambda J_n(\mu a)J'_m(\lambda a)], \quad (35.3)$$

if $n>-1$, $m>-1$.

Putting $m=n$ in this result we find that if $\lambda\neq\mu$,

$$\int_0^a xJ_n(\lambda x)J_n(\mu x)\,\mathrm{d}x=\frac{a}{\lambda^2-\mu^2}[\mu J_n(\lambda a)J'_n(\mu a)-\lambda J_n(\mu a)J'_n(\lambda a)]. \quad (35.4)$$

The corresponding expression for $\lambda=\mu$ is obtained by putting $\mu=\lambda+\varepsilon$, where ε is small, using Taylor's theorem and then letting ε tend to zero. We find that

$$\int_0^a x\{J_n(\lambda x)\}^2\,\mathrm{d}x=\tfrac{1}{2}a^2\left[\{J'_n(\lambda a)\}^2+\left(1-\frac{n^2}{\lambda^2a^2}\right)\right]\{J_n(\lambda a)\}^2. \quad (35.5)$$

Suppose now that λ and μ are positive roots of the transcendental equation

$$hJ_n(\lambda a)+k\lambda aJ'_n(\lambda a)=0, \quad (35.6)$$

where h and k are constants. It follows then that

$$\int_0^a xJ_n(\lambda x)J_n(\mu x)\,\mathrm{d}x=c_\lambda\delta_{\lambda,\mu}, \quad (35.7)$$

where

$$c_\lambda=\frac{\{J_n(\lambda a)\}^2}{2k^2\lambda^2}\{k^2\lambda^2a^2+h^2-k^2n^2\}. \quad (35.8)$$

If we now suppose that we can expand an arbitrary function $f(x)$ in the form

$$f(x)=\sum_i a_iJ_n(\lambda_ix), \quad (35.9)$$

where the sum is taken over the positive roots of the equation (35.6), then we can determine the coefficients a_i as follows: Multiply both sides of equation (35.9) by $xJ_n(\lambda_jx)$ and integrate with respect to x from 0 to a; then

$$\int_0^a xf(x)J_n(\lambda_j x)\,dx = \sum_i a_i \int_0^a xJ_n(\lambda_i x)J_n(\lambda_j x)\,dx,$$

from which it follows that

$$a_j = \frac{1}{c_\lambda}\int_0^a xf(x)J_n(\lambda_j x)\,dx. \tag{35.10}$$

Because of its similarity to a Fourier series, a series of the type (35.9) is called a Fourier–Bessel series.

In particular if the sum is taken over the roots of the equation

$$J_n'(\lambda a) = 0 \tag{35.11}$$

then the coefficients of the sum (35.9) are given by

$$a_j = \frac{2\lambda_j^2}{\{J_n(\lambda_j a)\}^2}\frac{1}{(\lambda_j^2 a^2 - n^2)}\int_0^a xf(x)J_n(\lambda_j x)\,dx. \tag{35.12}$$

Similarly if the sum is taken over the positive roots of the equation

$$J_n(\lambda a) = 0, \tag{35.13}$$

we find that the coefficients a_j are given by the formula

$$a_j = \frac{2}{a^2\{J_n'(\lambda_j a)\}^2}\int_0^a xf(x)J_n(\lambda_j x)\,dx. \tag{35.14}$$

In this section no attempt has been made to discuss the very difficult problem of the convergence of Fourier–Bessel series. For a very full discussion of this topic the reader is referred to Chapter XVIII of Watson (1944).

Some properties of the zeros of Bessel functions are developed in Problem 4.16 below; these are of importance in applications of Bessel functions.

§36. The use of Bessel functions in potential theory

As an example of the use of Bessel functions in potential theory we shall consider the problem of determining a function $\psi(\rho, z)$

for the half-space $a \geqslant \rho \geqslant 0$, $z \geqslant 0$ satisfying the differential equation

$$\frac{\partial^2 \psi}{\partial \rho^2}+\frac{1}{\rho}\frac{\partial \psi}{\partial \rho}+\frac{\partial^2 \psi}{\partial z^2}=0 \tag{36.1}$$

and the boundary conditions:

(i) $\psi = f(\rho)$, on $z=0$;

(ii) $\psi \to 0$ as $z \to \infty$;

(iii) $\dfrac{\partial \psi}{\partial \rho}+\kappa\psi=0$ on $\rho=a$;

(iv) ψ remains finite as $\rho \to 0$.

We saw in §1 that a function of the form $\psi = R(\rho)Z(z)$ is a solution of equation (36.1) provided that

$$\frac{d^2 Z}{dz^2}-\lambda_i^2 Z=0 \tag{36.2}$$

and that

$$\frac{d^2 R}{d\rho^2}+\frac{1}{\rho}\frac{dR}{d\rho}+\lambda_i^2 R=0 \tag{36.3}$$

where λ_i is a constant of separation. To satisfy the boundary condition (ii) we must take solutions of equation (36.2) of the form

$$Z=e^{-\lambda_i z}$$

and to satisfy the condition (iv) we must take as the solutions of equation (36.3) functions of the form

$$R=J_0(\lambda_i \rho),$$

since the second solutions $Y_0(\lambda_i \rho)$ would become infinite in the region of the axis $\rho=0$.

The differential equation (36.1) and the boundary conditions (ii) and (iv) are satisfied by any sums of the form

$$\psi(\rho, z)=\sum_i a_i e^{-\lambda_i z} J_0(\lambda_i \rho), \tag{36.4}$$

where the a_i and λ_i are constants. But if we are to satisfy the boundary condition (iii) we must take the sum over the positive roots of the equation

$$\lambda_i J_0'(\lambda_i a)+\kappa J_0(\lambda_i a)=0. \tag{36.5}$$

The solution is determined therefore if we can find constants a_i such that condition (i) is satisfied, i.e. such that

$$f(\rho)=\sum_i a_i J_0(\lambda_i\rho). \tag{36.6}$$

From equations (35.10) and (35.8) we see that we must take

$$a_i=\frac{2\lambda_i^2}{a^2(\lambda_i^2+\kappa^2)\{J_0(\lambda_i a)\}^2}\int_0^a \rho' f(\rho')J_0(\lambda_i\rho')\,d\rho'. \tag{36.7}$$

Hence the required solution is

$$\psi(\rho,z)=\frac{2}{a^2}\sum_i \frac{\lambda_i^2 e^{-\lambda_i z}J_0(\lambda_i\rho)}{(\lambda_i^2+\kappa^2)\{J_0(\lambda_i a)\}^2}\int_0^a \rho' f(\rho')J_0(\lambda_i\rho')\,d\rho', \tag{36.8}$$

where the sum is taken over the positive roots of the equation (36.5).

If, instead of the boundary condition (iii), we had the condition $\psi=0$ on $\rho=a$, then it is easily seen that the solution would have been

$$\psi(\rho,z)=\frac{2}{a^2}\sum_i \frac{e^{-\lambda_i z}J_0(\lambda_i\rho)}{\{J_1(\lambda_i a)\}^2}\int_0^a \rho' f(\rho')J_0(\lambda_i\rho')\,d\rho', \tag{36.9}$$

where the sum is taken over the positive roots λ_i of the transcendental equation

$$J_0(\lambda a)=0. \tag{36.10}$$

For example suppose that ψ satisfies the conditions $\psi=0$ on $\rho=a$, $\psi\to 0$ as $z\to\infty$, $\psi=\psi_0(a^2-\rho^2)$ on $z=0$, $0\leqslant\rho\leqslant a$, then the solution to the problem is given by equation (36.9) with $f(\rho')=\psi_0(a^2-\rho'^2)$. By equation (32.5) we have

$$\begin{aligned}\int_0^a \rho' f(\rho')J_0(\lambda_i\rho')\,d\rho' &= \frac{4a\psi_0}{\lambda_i^3}J_1(\lambda_i a)-\frac{2a^2\psi}{\lambda_i^2}J_0(\lambda_i a)\\ &= \frac{4a\psi_0}{\lambda_i^3}J_1(\lambda_i a),\end{aligned}$$

since λ_i is a root of equation (36.10). Thus the required solution is

$$\psi(\rho,z)=\frac{8\psi_0}{a}\sum_i \frac{e^{-\lambda_i z}J_0(\lambda_i\rho)}{\lambda_i^3 J_1(\lambda_i a)}. \tag{36.11}$$

Tables of the first forty zeros ξ_i of the function $J_0(\xi)$ with the corresponding values of $J_1(\xi_i)$ are available in Watson (1944) so that it is convenient to express results of the kind (36.11) in terms of them. It is readily seen that in this case

$$\psi(\rho, z)=8a^2\psi_0\sum_i \frac{e^{-\xi_i\zeta}J_0(\alpha\xi_i)}{\xi_i^3 J_1(\xi_1)}, \tag{36.12}$$

where $\zeta = z/a$ and $\alpha = \rho/a$.

§37. Asymptotic expansions of Bessel functions

In certain physical problems it is desirable to know the value of a Bessel function for large values of its argument. In this section we shall derive the asymptotic expansion of the Bessel function $J_n(x)$ of the first kind and merely indicate the results for the other Bessel functions occurring in mathematical physics.

It is easily shown that $x^{-n}J_n(x)$ satisfies the differential equation

$$xy''(x)+(2n+1)y'(x)+xy(x)=0.$$

From equation (3.7) we see that

$$x^{-n}J_n(x)=\frac{1}{2^n\Gamma(\frac{1}{2})\Gamma(n+\frac{1}{2})}\int_C (1-s^2)^{n-\frac{1}{2}}e^{isx}\,ds,$$

the multiplicative constant being chosen to give the correct value to the constant

$$\lim_{x\to 0} x^{-n}J_n(x)$$

and the contour C chosen such that

$$\Delta_C[(s^2-1)^{n+\frac{1}{2}}e^{isx}]=0.$$

From this result we can easily deduce by deforming C that

$$J_n(x)=\frac{(\frac{1}{2}x)^n}{\Gamma(\frac{1}{2})\Gamma(n+\frac{1}{2})}\left\{\int_{L_1} (1-t^2)^{n-\frac{1}{2}}e^{ixt}\,dt+\int_{L_2} (1-t^2)^{n-\frac{1}{2}}e^{ixt}\,dt\right\}, \tag{37.1}$$

where L_1 is the straight line $\mathscr{R}(t)=-1$ in the upper half of the complex t-plane and L_2 is the corresponding part of the straight line $\mathscr{R}(t)=+1$. By changing the variable from t to $u=ix(1-t)$ in the first integral and to $u=-ix(1-t)$ in the second we see that

$$J_n(x)=\sqrt{\frac{2}{\pi x}}\,\{j_n(x)+j_n^*(x)\} \tag{37.2}$$

where

$$j_n(x)=\frac{1}{2\Gamma(n+\frac{1}{2})}\exp\{ix-(\tfrac{1}{2}n+\tfrac{1}{4})\pi i\}\int_0^\infty e^{-u}u^{n-\frac{1}{2}}\left(1+\frac{iu}{2x}\right)^{n-\frac{1}{2}}du$$

and $j_n^*(x)$ denotes its complex conjugate. Expanding $(1+iu/2x)^{n-\frac{1}{2}}$ by the binomial theorem and integrating term by term we find that

$$j_n(x)=\tfrac{1}{2}\exp\{ix-(\tfrac{1}{2}n+\tfrac{1}{4})\pi_i\}\,{}_2F_0\left(\tfrac{1}{2}+n,\tfrac{1}{2}-n;\frac{1}{2ix}\right). \tag{37.3}$$

If we adopt Hankel's convention of writing

$$(n,r)=(-1)^r\frac{(\frac{1}{2}-n)_r(\frac{1}{2}+n)_r}{r!}$$

in equation (37.3) and substitute the result in equation (37.2), we find that for large values of x the asymptotic expansion of the Bessel function $J_n(x)$ is

$$J_n(x)\sim\sqrt{\frac{2}{\pi x}}\left\{\cos(x-\tfrac{1}{2}n\pi-\tfrac{1}{4}\pi)\sum_{r=0}^{\infty}\frac{(-1)^r(n,2r)}{(2x)^{2r}}\right.$$
$$\left.-\sin(x-\tfrac{1}{2}n\pi-\tfrac{1}{4}\pi)\sum_{r=0}^{\infty}\frac{(-1)^r(n,2r+1)}{(2x)^{2r+1}}\right\}. \tag{37.4}$$

The corresponding expansion for the Bessel function of the second kind is found to be

$$Y_n(x)\sim\sqrt{\frac{2}{\pi x}}\left\{\sin(x-\tfrac{1}{2}n\pi-\tfrac{1}{4}\pi)\sum_{r=0}^{\infty}\frac{(-1)^r(n,2r)}{(2x)^{2r}}\right.$$
$$\left.+\cos(x-\tfrac{1}{2}n\pi-\tfrac{1}{4}\pi)\sum_{r=0}^{\infty}\frac{(-1)^r(n,2r+1)}{(2x)^{2r+1}}\right\}. \tag{37.5}$$

Substituting these asymptotic expressions in equations (30.18) and (30.19) we find that as $x\to\infty$,

$$H_n^{(1)}(x)\sim\left(\frac{8}{\pi x}\right)^{\frac{1}{2}}j_n(x),\qquad H_n^{(2)}(x)\sim\left(\frac{8}{\pi x}\right)^{\frac{1}{2}}j_n^*(x) \tag{37.6}$$

where $j_n(x)$ is given by equation (37.3).

In certain problems only a very crude approximation to the behaviour of the Bessel function is desired. In these circumstances the following formulae are usually sufficient:

$$J_n(x) \sim \sqrt{\frac{2}{\pi x}} \cos(x - \tfrac{1}{2}n\pi - \tfrac{1}{4}\pi),$$
$$Y_n(x) \sim \sqrt{\frac{2}{\pi x}} \sin(x - \tfrac{1}{2}n\pi - \tfrac{1}{4}\pi); \tag{37.7}$$

$$H_n^{(1)}(x) \sim \sqrt{\frac{2}{\pi x}} \exp(ix - \tfrac{1}{2}n\pi i - \tfrac{1}{4}\pi i),$$
$$H_n^{(2)}(x) \sim \sqrt{\frac{2}{\pi x}} \exp(-ix + \tfrac{1}{2}n\pi i - \tfrac{1}{4}\pi i). \tag{37.8}$$

Similar formulae exist for the modified Bessel functions. Proceeding in the same way as in the establishment of equation (28.1) we can show that

$$I_n(x) = \frac{1}{\sqrt{\pi}\,\Gamma(n+\frac{1}{2})} (\tfrac{1}{2}x)^n \int_C e^{\pm xt}(1-t^2)^{n-\frac{1}{2}}\, dt,$$

which becomes

$$I_n(x) = \frac{1}{\sqrt{(2\pi x)}\,\Gamma(n+\frac{1}{2})} \left\{\exp\{-x-(n+\tfrac{1}{2})\pi\} \int_0^\infty e^{-u}u^{n-\frac{1}{2}}\left(1+\frac{u}{2x}\right)^{n-\frac{1}{2}} du \right.$$
$$\left. + e^x \int_0^\infty e^{-u}u^{n-\frac{1}{2}}\left(1-\frac{u}{2x}\right)^{n-\frac{1}{2}} du\right\}$$

by a simple change of variable. By a method similar to that employed above to obtain the asymptotic expansion of $J_n(x)$, we can then show that if $-\frac{1}{2}\pi < \arg x < \frac{3}{2}\pi$,

$$I_n(x) \sim \frac{1}{\sqrt{(2\pi x)}} e^x \sum_{r=0}^\infty \frac{(-1)^r(n,r)}{(2x)^r} + \frac{\exp\{-x+(n+\frac{1}{2})\pi i\}}{\sqrt{(2\pi x)}} \sum_{r=0}^\infty \frac{(n,r)}{(2x)^r}$$

and that if $-\frac{3}{2}\pi < \arg x < \frac{1}{2}\pi$ the factor $\exp\{-x+(n+\frac{1}{2})\pi i\}$ is replaced by $\exp\{-x-(n+\frac{1}{2})\pi i\}$. The corresponding formula for the modified Bessel of the second kind is

$$K_n(x) \sim \left(\frac{\pi}{2x}\right)^{\frac{1}{2}} e^{-x} \sum_{r=0}^\infty \frac{(n,r)}{(2x)^r}$$

as $x \to \infty$.

Problems IV

4.1 Making use of Problem 2.2 and of the expansion (25.8), expand $\cos(x\sin\theta)$ as a power series in $\sin\theta$ in two ways. Hence by equating powers of $\sin^{2s}\theta$ show that if s is a positive integer,

$$x^{2s}=2^{2s+1}\sum_{n=s}^{\infty}\frac{(n+s-1)!}{(n-s)!}J_{2n}(x).$$

Derive the corresponding result for x^{2s+1} and show that the two results may be combined into the single formula

$$x^r=2^r\sum_{n=0}^{\infty}\frac{(r+n-1)!}{n!}(r+2n)J_{r+2n}(x)$$

$(r=1,2,3,\ldots)$.

4.2 Show that if u and ζ denote the eccentric and mean anomalies of a planet, then

$$\cos(nu)=n\sum_{m=-\infty}^{\infty}\frac{1}{m}J_{m-n}(me)\cos(m\zeta),$$

$$\sin(nu)=n\sum_{m=-\infty}^{\infty}\frac{1}{m}J_{m-n}(me)\sin(m\zeta).$$

4.3 Making use of the expansion for $P_n(\cos\theta)$ given in Problem 3.19, show that

$$\sum_{n=0}^{\infty}\frac{r^n}{n!}P_n(\cos\theta)=e^{r\cos\theta}J_0(r\sin\theta).$$

4.4 Show that

(i) $8J_n'''(z)=J_{n-3}(z)-3J_{n-1}(z)+3J_{n+1}(z)-J_{n+3}(z)$;
(ii) $4J_0'''(z)+3J_0'(z)+J_3(z)=0$.

4.5 Prove that

$$\tfrac{1}{2}+\sum_{r=1}^{N}(-1)^rJ_0(rx)=\frac{(-1)^N}{\pi}\int_0^x\frac{\cos(N+\frac{1}{2})u}{\cos\frac{1}{2}u}\frac{du}{\sqrt{(x^2-u^2)}}$$

and deduce that

$$\tfrac{1}{2}+\sum_{r=1}^{\infty}(-1)^rJ_0(rx)=0.$$

4.6 Show that the curve with parametric equations

$$x=t-\sin t,\qquad y=\cos t/(1-\cos t)$$

may be represented in the interval $0<t<\pi$ by the Fourier series

$$y=2\sum_{n=1}^{\infty} J_n(n)\cos(nx).$$

4.7 Prove that

$$e^{ikr\cos\theta}=\left(\frac{\pi}{2kr}\right)^{\frac{1}{2}}\sum_{n=0}^{\infty}(2n+1)e^{\frac{1}{2}n\pi i}J_{n+\frac{1}{2}}(kr)P_n(\cos\theta).$$

4.8 Prove that

$$\lim_{n\to\infty} P_n\left(\cos\frac{x}{n}\right)=J_0(x).$$

4.9 Show that

(i) $\displaystyle\sum_{n=0}^{\infty}\frac{x^n}{n!}J_n(a)=J_0\{\sqrt{(a^2-2ax)}\};$

(ii) $\displaystyle\sum_{n=0}^{\infty}\frac{(-x)^n}{n!}a^{-\frac{1}{2}m-\frac{1}{2}n}J_{m+n}(2\sqrt{a})=(x+a)^{-\frac{1}{2}m}J_m\{2\sqrt{(x+a)}\}.$

4.10 Prove that if n is a positive integer,

$$J_n(2\sqrt{x})=(-1)^n x^{\frac{1}{2}n}\frac{d^n}{dx^n}J_0(2\sqrt{x}).$$

4.11 Show that if $x>a$,

$$\int_0^{\pi} e^{a\cos\theta}\cos(x\sin\theta)\,d\theta=J_0\{\sqrt{(x^2-a^2)}\}.$$

4.12 Prove that if $-1<x<1$,

$$\frac{1}{\sqrt{(1-x^2)}}=\tfrac{1}{2}\pi+\pi\sum_{m=1}^{\infty}J_0(m\pi)\cos(m\pi x).$$

Deduce that

$$J_0(x)=\frac{\sin x}{x}\left\{1+2x^2\sum_{m=1}^{\infty}\frac{(-1)^m J_0(m\pi)}{x^2-m^2\pi^2}\right\}.$$

4.13 Prove that

$$J'_\nu(x)J_{-\nu}(x)-J_\nu(x)J'_{-\nu}(x)\equiv\frac{A}{x},$$

where A is a constant, and, by considering the series for $J_\nu(x)$ and $J_{-\nu}(x)$ when x is small, show that $A=(2/\pi)\sin(\nu\pi)$.

4.14 Show that the complete solution of Bessel's equation may be written in the form

$$AJ_n(x)+BJ_n(x)\int^x \frac{d\xi}{\xi\{J_n(\xi)\}^2},$$

where A and B are arbitrary constants.

4.15 Show that the complete solution of the differential equation

$$\frac{d^2y}{dx^2}+\tfrac{1}{3}xy=0$$

is

$$y=\sqrt{x}\{AJ_{\frac{1}{3}}(\xi)+BJ_{-\frac{1}{3}}(\xi)\},$$

where $\xi^2=4x^3/27$ and A and B are arbitrary constants.

4.16 Show that if a and b are real constants, then the roots of the equation

$$axJ_n'(x)+bJ_n(x)=0$$

are simple roots except possibly the root $x=0$.

Show also that the equations $J_n(x)=0$, $J_n'(x)=0$ have no roots in common except possibly $x=0$.

4.17 Deduce from equation (32.10) that

(i) $\displaystyle\int_0^x J_0(tx)\cos(ax)\,dx=(t^2-a^2)^{-\frac{1}{2}}H(t-a),$

(ii) $\displaystyle\int_0^\infty J_0(tx)\sin(ax)\,dx=(a^2-t^2)^{-\frac{1}{2}}H(a-t),$

and hence show that

(iii) $\displaystyle\int_0^\infty x^{-1}\sin(ax)J_0(tx)\,dx=\begin{cases}\frac{1}{2}\pi, & 0<t<a,\\ \sin^{-1}(a/t), & 0<a<t;\end{cases}$

(iv) $\displaystyle\int_0^\infty \sin(ax)J_1(tx)\,dx=at^{-1}(t^2-a^2)^{-\frac{1}{2}}H(t-a).$

4.18 Prove that if $\mu>-\frac{1}{2}$, $\nu>-1$,

$$\int_0^x u^{\mu+1}(x^2-u^2)^\nu J_\mu(uy)\,du=2^\nu x^{\mu+\nu+1}y^{-\nu-1}\Gamma(\nu+1)J_{\mu+\nu+1}(xy).$$

Deduce that

$$\int_0^\infty y^{-\nu}J_\mu(uy)J_{\nu+\mu+1}(xy)\,dy=\frac{2^{-\nu}}{\Gamma(\nu+1)}x^{-\mu-\nu-1}u^\mu(x^2-u^2)^\nu H(x-u).$$

4.19 Prove that

$$V(\rho, z)=\int_0^\infty tg(t)e^{-tz}J_0(t\rho)\,dt$$

satisfies Laplace's equation in the half-space $z \geqslant 0$ and the condition $V(\rho, z)\to 0$ as $(\rho^2+z^2)^{\frac{1}{2}}\to\infty$. If, in addition,

$$V(\rho, 0)=f(\rho), \qquad \rho\geqslant 0,$$

show that

$$g(t)=\int_0^\infty \rho f(\rho)J_0(t\rho)\,d\rho$$

4.20 Show that

$$V(\rho, z)=\frac{2}{\pi}\int_0^\infty t^{-1}\sin(at)e^{-zt}J_0(t\rho)\,dt$$

is the solution of Laplace's equation in the half-space $z>0$, which tends to zero as $(\rho^2+z^2)^{\frac{1}{2}}\to\infty$, and which satisfies the mixed boundary conditions

$$V(\rho, 0)=1, \qquad 0<\rho<a,$$
$$\frac{\partial V(\rho, 0)}{\partial z}=0, \qquad \rho>a.$$

(This boundary value problem arises in the analysis of the problem of the electrified disk in electrostatics.)

4.21 If $x>1$ and $m+n+1>0$, prove that

$$\int_0^\infty e^{-xt}I_{n+\frac{1}{2}}(t)t^{m-\frac{1}{2}}\,dt=\sqrt{\frac{2}{\pi}}(x^2-1)^{-\frac{1}{2}m}Q_n^m(x),$$

where $Q_n^m(x)$ denotes the associated Legendre function of the second kind.

4.22 Prove that

$$I_n(x+y)=\sum_{m=0}^{n} I_m(x)I_{n-m}(y)+\sum_{m=1}^{\infty}\{I_m(x)I_{n+m}(y)+I_{n+m}(x)I_m(y)\}.$$

4.23 Show that

$$\int_{-1}^{1} e^{x\mu} P_n(\mu)\,d\mu = \sqrt{\frac{2\pi}{x}}\, I_{n+\frac{1}{2}}(x).$$

4.24 Prove that

$$\sum_{n=-\infty}^{\infty} J_n(kx)t^n = \exp\left\{-\frac{x}{2t}\left(k-\frac{1}{k}\right)\right\} \sum_{n=-\infty}^{\infty} k^n t^n J_n(x)$$

and deduce that

(i) $J_n(re^{i\theta}) = \sum_{m=0}^{\infty} J_{n+m}(r)e^{i(n+m)\theta} \frac{(-ir\sin\theta)^m}{m!}$;

(ii) $I_n(x) = \sum_{m=0}^{\infty} \frac{x^m}{m!} J_{n+m}(x)$.

4.25 Using the expansion of Problem 4.24 prove that $J_r(ae^{i\alpha} + be^{i\beta})$ is the coefficient of t^r in the expansion of

$$\exp\left\{-\frac{i(a\sin\alpha + b\sin\beta)}{t}\right\} \sum_{m=-\infty}^{\infty} \sum_{n=-\infty}^{\infty} e^{in\alpha - im\beta} J_n(a)J_m(b)t^{n+m}.$$

By putting

$$R = a\cos\alpha + b\cos\beta, \quad 0 = a\sin\alpha + b\sin\beta, \quad \beta = \alpha + \theta - \pi$$

prove Neumann's addition theorem

$$J_n(R) = \left(\frac{a - be^{-i\theta}}{a - be^{i\theta}}\right)^{\frac{1}{2}n} \sum_{m=-\infty}^{\infty} J_{n+m}(a)J_m(b)e^{im\theta},$$

where $R^2 = a^2 + b^2 - 2ab\cos\theta$.

4.26 Prove that

$$J_0(ax)J_0(bx) = \frac{1}{\pi}\int_0^{\pi} J_0(Rx)\,d\theta,$$

where $R^2 = a^2 + b^2 - 2ab\cos\theta$, and deduce that

$$\int_0^{\infty} J_0(ax)J_0(bx)e^{-cx}\,dx = \frac{k}{\pi\sqrt{(ab)}} K(k)$$

where

$$K(k) = \int_0^{\frac{1}{2}\pi} \frac{d\phi}{\sqrt{(1-k^2\sin^2\phi)}}, \qquad k^2 = \frac{4ab}{(a+b)^2 + c^2}.$$

Prove, in a similar way, that

$$\int_0^\infty J_1(ax)J_1(bx)e^{-cx}\,dx = \frac{2}{\pi k\sqrt{(ab)}}\{(1-\tfrac{1}{2}k^2)K(k)-E(k)\}$$

where k and $K(k)$ are as defined above and

$$E(k) = \int_0^{\frac{1}{2}\pi} \sqrt{(1-k^2\sin^2\phi)}\,d\phi.$$

4.27 Show that

$$J_m(x)J_n(x) = \frac{1}{\pi}\exp\{\tfrac{1}{2}(n-m)\pi i\}\int_0^\pi J_{n-m}(2x\sin\theta)e^{(n+m)\theta i}\,d\theta$$

and hence that

$$\{J_n(x)\}^2 = \frac{1}{\pi}\int_0^\pi J_{2n}(2x\sin\theta)\,d\theta.$$

Deduce that

$$\{J_n(x)\}^2 = (\tfrac{1}{2}x)^{2n}\sum_{s=0}^{\infty}\frac{(2n+2s)!}{s!(2n+s)!(n+s)!}(-\tfrac{1}{4}x^2)^s.$$

4.28 Prove that

$$\int_0^\infty J_0(ax)J_0(bx)e^{-p^2x^2}x\,dx = \frac{1}{2p}\exp\left(-\frac{a^2-b^2}{4p^2}\right)I_0\left(\frac{ab}{2p}\right).$$

4.29 We define the **Bessel-integral function of order n** by the equation

$$Ji_n(x) = -\int_0^x \frac{J_n(u)}{u}\,du, \qquad n=1,2,3,\ldots$$

Prove that if n is even

$$Ji_n(x) = \frac{1}{\pi}\int_0^\pi \cos(n\theta)\mathrm{Ci}(x\sin\theta)\,d\theta,$$

where $\mathrm{Ci}(x)$ denotes the cosine integral, and derive the corresponding expression when n is odd.

Show that:

(i) $\mathrm{li}\{\exp[\tfrac{1}{2}x(t-1/t)]\} = \sum_{n=-\infty}^{\infty} t^n Ji_n(x);$

(ii) $Ji_n'(x) = J_n(x)/x;$

(iii) $(n-1)Ji_{n-1}(x)-(n+1)Ji_{n+1}(x) = 2nJi_n'(x);$

(iv) $\mathrm{Ci}(x) = Ji_0(x) - 2Ji_2(x) + 2Ji_4(x) - \ldots;$
(v) $\mathrm{Si}(x) = 2Ji_1(x) - 2Ji_3(x) + 2Ji_5(x) - \ldots;$
(vi) $Ji_0(x) = \gamma + \log(\tfrac{1}{2}x) - \dfrac{x^2}{8}\,{}_2F_3(1, 1; 2, 2, 2; -x^2/4).$

4.30 If

$$I_m = \frac{1}{J_1(\lambda)} \int_0^1 x^{2m+1} J_0(\lambda x)\,dx$$

where λ is a zero of $J_0(x)$, show that

$$I_m = \frac{1}{\lambda}\left[1 - \frac{4m^2}{\lambda^2} + \frac{4^2m^2(m-1)^2}{\lambda^4} + \ldots + (-1)^m \frac{4^m m^2(m-1)^2 \ldots 2^2 . 1^2}{\lambda^{2m}}\right].$$

Show that

$$1 - x^{2m} = \sum_{i=1}^{\infty} a_i J_0(\lambda_i x), \qquad 0 \leq x \leq 1,$$

where $\lambda_1, \lambda_2, \ldots$ are the positive zeros of $J_0(x)$ and

$$a_i = \frac{2}{J_1(\lambda_i)}(1 - I_m).$$

Putting $m = 1, 2, 3, \ldots$, and rearranging the series, show that

$$\sum_{i=1}^{\infty} \frac{J_0(\lambda_i x)}{\lambda_i^3 J_1(\lambda_i)} = \tfrac{1}{8}(1-x)^2;$$

$$\sum_{i=1}^{\infty} \frac{J_0(\lambda_i x)}{\lambda_i^5 J_1(\lambda_i)} = \tfrac{1}{128}(1-x^2)(3-x^2);$$

$$\sum_{i=1}^{\infty} \frac{J_0(\lambda_i x)}{\lambda_i^7 J_1(\lambda_i)} = \tfrac{1}{4608}(1-x^2)(19-8x^2+x^4).$$

Deduce that

$$\sum_{i=1}^{\infty} \frac{1}{\lambda_i^3 J_1(\lambda_i)} = \frac{1}{8}; \quad \sum_{i=1}^{\infty} \frac{1}{\lambda_i^5 J_1(\lambda_i)} = \frac{3}{128}; \quad \sum_{i=1}^{\infty} \frac{1}{\lambda_i^7 J_1(\lambda_i)} = \frac{19}{4608}.$$

4.31 Let $\alpha_1, \alpha_2, \ldots$ denote the positive zeros of $J_\nu(x)$. Show that

$$\sum_{i=1}^{\infty} \frac{1}{\alpha_i J_{\nu+1}(\alpha_i)} \int_0^1 x f(x) J_\nu(\alpha_i x)\,dx = \tfrac{1}{2} \int_0^1 x^{\nu+1} f(x)\,dx.$$

Taking $f(x)=x^\nu$, $x^{\nu+2}$ show that

$$\sum_{i=1}^{\infty}\frac{1}{\alpha_i^2}=\frac{1}{4(\nu+1)};\qquad \sum_{i=1}^{\infty}\frac{1}{\alpha_i^4}=\frac{1}{16(\nu+1)^2(\nu+2)}.$$

4.32 Putting $\nu=0$, $f(x)=x^m$ in Problem 4.31 show that

$$\sum_{i=1}^{\infty}\frac{I_m}{\lambda_i}=\frac{1}{4(m+1)}$$

where $\lambda_1, \lambda_2, \ldots$ are the positive zeros of $J_0(x)$ and I_m is defined in (26).

Substituting the value for I_m derived in Problem 4.30 and using the fact (obtained by putting $\nu=0$ in Problem 4.31) that $\sum \lambda_i^{-2}=\frac{1}{4}$, show that the sum

$$S_{2m}=\sum_{i=1}^{\infty}\frac{1}{\lambda_i^{2m}}$$

satisfies the recurrence formula

$$S_{2m+2}=\sum_{r=1}^{m}(-1)^{m-r}(\tfrac{1}{4})^{m-r+1}\frac{S_{2r}}{[(m-r+1)!]^2}+(-1)^m(\tfrac{1}{4})^{m+1}\frac{1}{m!(m+1)!}.$$

Deduce that

$$\sum_{i=1}^{\infty}\frac{1}{\lambda_i^4}=\frac{1}{32};\qquad \sum_{i=1}^{\infty}\frac{1}{\lambda_i^6}=\frac{1}{192};\qquad \sum_{i=1}^{\infty}\frac{1}{\lambda_i^8}=\frac{11}{12288}.$$

4.33 Show that

$$\sum_{i=1}^{\infty}\frac{J_0(\lambda_i x)}{\lambda_i^2 J_1^2(\lambda_i)}=-\tfrac{1}{2}\log x.$$

Multiplying both sides of this equation by x and integrating term by term, show that

$$\sum_{i=1}^{\infty}\frac{J_1(\lambda_i x)}{\lambda_i^3 J_1^2(\lambda_i)}=\tfrac{1}{8}x(1-2\log x),$$

and hence that

$$\sum_{i=1}^{\infty}\frac{1}{\lambda_i^3 J_1(\lambda_i)}=\frac{1}{8}.$$

5

The functions of Hermite and Laguerre

§38. The Hermite polynomials

The **Hermite polynomial** $H_n(x)$ is defined for integral values of n and all real values of x by the identity

$$e^{2tx-t^2} = \sum_{n=0}^{\infty} \frac{H_n(x)}{n!} t^n. \tag{38.1}$$

If we write

$$f(x, t) = e^{2tx-t^2} = e^{x^2} e^{-(x-t)^2},$$

then it follows from Taylor's theorem that

$$H_n(x) = \left(\frac{\partial^n f}{\partial t^n}\right)_{t=0} = e^{x^2}\left[\frac{\partial^n}{\partial t^n} e^{-(x-t)^2}\right]_{t=0}.$$

Now it is obvious from the form of the function $\exp\{-(x-t)^2\}$ that

$$\left[\frac{\partial^n}{\partial t^n} e^{-(x-t)^2}\right]_{t=0} = (-1)^n \frac{d^n}{dx^n}(e^{-x^2})$$

and so we have the form

$$H_n(x) = (-1)^n e^{x^2} \frac{d^n}{dx^n}(e^{-x^2}) \tag{38.2}$$

for the calculation of the polynomial $H_n(x)$.

It follows from this formula that the first eight Hermite polynomials are:

$$H_0(x) = 1,$$
$$H_1(x) = 2x,$$
$$H_2(x) = 4x^2 - 2,$$
$$H_3(x) = 8x^3 - 12x,$$
$$H_4(x) = 16x^4 - 48x^2 + 12,$$

$$H_5(x)=32x^5-160x^3+120x,$$
$$H_6(x)=64x^6-480x^4+720x^2-120,$$
$$H_7(x)=128x^7-1344x^5+3360x^3-1680x.$$

In general we have

$$H_n(x)=(2x)^n-\frac{n(n-1)}{1!}(2x)^{n-2}+\frac{n(n-1)(n-3)(n-4)}{2!}(2x)^{n-4}+\dots$$

or, in the notation of Section 12,

$$H_n(x)=(2x)^n\,{}_2F_0\left(-\tfrac{1}{2}n,\tfrac{1}{2}-\tfrac{1}{2}n;-\frac{1}{x^2}\right). \tag{38.3}$$

Recurrence formulae for the Hermite polynomials follow directly from the defining relation (38.1). If we differentiate both sides of that equation with respect to x we obtain the relation

$$2te^{2xt-t^2}=\sum_{n=0}^{\infty}\frac{H_n'(x)}{n!}t^n,$$

from which it follows directly that

$$2nH_{n-1}x=H_n'(x). \tag{38.4}$$

On the other hand if we differentiate both sides of the identity (38.1) with respect to t we obtain the relation

$$2(x-t)e^{2tx-t^2}=\sum_{n=1}^{\infty}\frac{H_n(x)}{(n-1)!}t^{n-1}$$

which can be written in the form

$$2x\sum_{n=0}^{\infty}\frac{H_n(x)}{n!}t^n-2\sum_{n=0}^{\infty}\frac{H_n(x)}{n!}t^{n+1}=\sum_{n=1}^{\infty}\frac{H_n(x)}{(n-1)!}t^{n-1},$$

to yield the identity

$$2xH_n(x)=2nH_{n-1}(x)+H_{n+1}(x) \tag{38.5}$$

by the identification of coefficients of t^n.

Eliminating $2nH_{n-1}(x)$ from equations (38.4) and (38.5) we obtain the relation

$$H_n'(x)=2xH_n(x)-H_{n+1}(x). \tag{38.6}$$

Differentiating both sides of this identity we find that

$$H_n''(x)=2xH_n'(x)+2H_n(x)-H_{n+1}'(x)$$

and, by equation (38.4), $H'_{n+1}(x) = 2(n+1)H_n(x)$, so that

$$H''_n(x) - 2xH'_n(x) + 2nH_n(x) = 0. \tag{38.7}$$

In other words $y = H_n(x)$ is a solution of the linear differential equation

$$y'' - 2xy' + 2ny = 0. \tag{38.8}$$

§39. Hermite's differential equation

We saw in the last section that $H_n(x)$ is a solution of the differential equation (38.8). Replacing the integer n in that equation by the parameter ν we obtain **Hermite's differential equation**

$$\frac{d^2y}{dx^2} - 2x\frac{dy}{dx} + 2\nu y = 0. \tag{39.1}$$

If we assume a solution of this equation in the form

$$y = \sum_{r=0}^{\infty} a_r x^{r+\rho}$$

and substitute in the equation (39.1), we obtain the recurrence relation

$$a_{r+2} = \frac{2(r+\rho-\nu)}{(r+\rho+2)(r+\rho+1)}\, a_r \tag{39.2}$$

on equating to zero the coefficient of $x^{r+\rho}$. Equating to zero the coefficient of $x^{\rho-2}$ we obtain the indicial equation

$$\rho(\rho-1) = 0. \tag{39.3}$$

Corresponding to the root $\rho = 0$ we have the recurrence relation

$$a_{r+2} = \frac{2(r-\nu)}{(r+1)(r+2)}\, a_r \tag{39.4}$$

which gives the solution

$$y_1(x) = a\left(1 - \frac{2\nu}{2!}x^2 + \frac{2^2\nu(\nu-2)}{4!}x^4 - \frac{2^3\nu(\nu-2)(\nu-4)}{6!}x^6 + \dots\right) \tag{39.5}$$

where a_1 is a constant.

Similarly, corresponding to the root $\rho=1$ of the indicial equation we have the recurrence relation

$$a_{r+2}=\frac{2(r+1-\nu)}{(r+3)(r+2)}a_r, \qquad (39.6)$$

from which is derived the solution

$$y_2(x)=a_2x\left(1-\frac{2(\nu-1)}{3!}x^2+\frac{2^2(\nu-1)(\nu-3)}{5!}x^4+\ldots\right), \qquad (39.7)$$

where a_2 is a constant. The general solution of Hermite's differential equation is therefore

$$y=y_1(x)+y_2(x). \qquad (39.8)$$

For general values of the parameter ν the two series for $y_1(x)$ and $y_2(x)$ are infinite. From equations (39.4) and (39.6) it follows that for both series

$$a_{r+2}\sim\frac{2}{r}a_r \quad \text{as} \quad r\to\infty. \qquad (39.9)$$

If we write

$$\exp(x^2)=b_0+b_2x^2+\ldots+b_rx^r+b_{r+2}x^{r+2}+\ldots$$

then

$$b_{r+2}\sim\frac{2}{r}b_r, \quad \text{as} \quad r\to\infty. \qquad (39.10)$$

Suppose that a_N/b_N is equal to a constant γ, which may be small or large, then it follows from equations (39.9) and (39.10) that, for large enough values of N, $a_{N+2m}/b_{N+2m}\sim\gamma$. In other words the higher terms of the series for $y_1(x)$, $y_2(x)$ differ from those of $\exp(x^2)$ only by multiplicative constants γ_1, γ_2, so that for large values of $|x|$,

$$y_1(x)\sim\gamma_1e^{x^2}, \qquad y_2(x)\sim\gamma_2e^{x^2}$$

since for such values the lower terms are unimportant.

We shall see later (§41 below), that in quantum mechanics we require solutions of Hermite's differential equation which do not become infinite more rapidly than $\exp(\frac{1}{2}x^2)$ as $|x|\to\infty$. It follows from the above considerations that such solutions are possible only if either $y_1(x)$ or $y_2(x)$ reduces to simply polynomials, and it is obvious from equations (39.5) and (39.7) that this occurs only if ν is a positive integer. For example, if ν is an *even* integer n we get the solution

$$y(x) = cH_n(x), \tag{39.11}$$

where c is a constant, by taking $a_2 = 0$,

$$a_1 = (-1)^{\frac{1}{2}n} \frac{n!}{(\frac{1}{2}n)!} c.$$

Similarly, if ν is an *odd* integer n we get the solution (39.11) by taking $a_1 = 0$ and

$$a_2 = (-1)^{\frac{1}{2}n - \frac{1}{2}} \frac{2n!}{(\frac{1}{2}n - \frac{1}{2})!} c.$$

Hermite's differential equation therefore possesses solutions which do not become infinite more rapidly than $\exp(\frac{1}{2}x^2)$ as $|x| \to \infty$ if and only if ν is a positive integer n. When this is so the required solution of Hermite's equation is given by equation (39.11).

The methods of §3 can, of course, be employed to derive contour integral solutions of Hermite's equation. The basic results are stated in Problem 5.3.

§40. Hermite functions

A differential equation closely related to Hermite's equation is

$$\frac{d^2\psi}{dx^2} + (\lambda - x^2)\psi = 0. \tag{40.1}$$

It is readily shown that if we transform the dependent variable from ψ to y, where

$$\psi = e^{-\frac{1}{2}x^2} y \tag{40.2}$$

and put $\lambda = 1 + 2\nu$, then y satisfies Hermite's equation (39.1). The general solution of equation (40.1) is therefore given by equations (40.2) and (39.8) with $y_1(x)$, $y_2(x)$ given by equations (39.5) and (39.7) respectively.

The argument at the end of the last section shows that the equation (40.1) possesses solutions which tend to zero as $|x| \to \infty$ if and only if the parameter λ is of the form $1 + 2n$, where n is a positive integer. When λ is of this form the required solution of

(40.1) is a constant multiple of the function $\Psi_n(x)$ defined by the equation

$$\Psi_n(x) = e^{-\frac{1}{2}x^2} H_n(x), \tag{40.3}$$

where $H_n(x)$ is the Hermite polynomial of degree n. The function $\Psi_n(x)$ is called a **Hermite function of order *n*.**

The recurrence relations for $\Psi_n(x)$ follow immediately from those for $H_n(x)$. For instance equation (38.4) is equivalent to the relation

$$2n\Psi_{n-1}(x) = x\Psi_n(x) + \Psi_n'(x), \tag{40.4}$$

and equation (38.5) is unaltered in form so that

$$2x\Psi_n(x) = 2n\Psi_{n-1}(x) + \Psi_{n+1}(x). \tag{40.5}$$

Eliminating $2n\Psi_{n-1}(x)$ from equations (40.4) and (40.5) we have the relation

$$\Psi_n'(x) = x\Psi_n(x) - \Psi_{n+1}(x). \tag{40.6}$$

From the point of view of mathematical physics the most important properties of Hermite functions concern integrals involving products of two of them. In establishing most of these properties the starting point is the observation that the function $\Psi_n(x)$ satisfies the relation

$$\Psi_n'' + (2n+1-x^2)\Psi_n = 0, \tag{40.7}$$

as is obvious merely by substituting $2n+1$ for λ in equation (40.1).

Writing down the corresponding relation for Ψ_m,

$$\Psi_m'' + (2m+1-x^2)\Psi_m = 0, \tag{40.8}$$

multiplying it by Ψ_n and subtracting it from equation (40.7) multiplied by Ψ_m we obtain, as a result of integrating over $(-\infty, \infty)$, the relation

$$2(m-n)\int_{-\infty}^{\infty} \Psi_m\Psi_n\, dx = \int_{-\infty}^{\infty} (\Psi_m\Psi_n'' - \Psi_n\Psi_m)\, dx.$$

Now an integration by parts shows that the right-hand side of this equation has the value

$$[\Psi_m\Psi_n' - \Psi_n\Psi_m']_{-\infty}^{\infty} - \int_{-\infty}^{\infty} (\Psi_m'\Psi_n' - \Psi_n'\Psi_m')\, dx$$

and, if we remember that, for all positive integers n, $\Psi_n(x) \to 0$ as $|x| \to \infty$, we see that this has the value zero. Hence if we let

$$I_{m,n} = \int_{-\infty}^{\infty} \Psi_m(x)\Psi_n(x)\,\mathrm{d}x,$$

we see that

$$I_{m,n} = 0, \quad \text{if} \quad m \neq n. \tag{40.9}$$

In particular

$$I_{n-1,n+1} = 0$$

so that from equation (40.5) we have

$$\int_{-\infty}^{\infty} 2x\Psi_n(x)\Psi_{n-1}(x)\,\mathrm{d}x = 2nI_{n-1,n-1}. \tag{40.10}$$

Now if in equation (40.3) we substitute for $H_n(x)$ from equation (38.2), we have

$$\Psi_n(x) = (-1)^n e^{\frac{1}{2}x^3} \frac{\mathrm{d}^n}{\mathrm{d}x^n}(e^{-x^2}), \tag{40.11}$$

so that the left-hand side of equation (40.10) is equal to

$$-\int_{-\infty}^{\infty} 2xe^{x^2} \frac{\mathrm{d}^n}{\mathrm{d}x^n}(e^{-x^2}) \frac{\mathrm{d}^{n-1}}{\mathrm{d}x^{n-1}}(e^{-x^2})\,\mathrm{d}x$$

and an integration by parts shows that this is equal to

$$I_{n,n} + I_{n+1,n-1}$$

i.e. to $I_{n,n}$. Hence from equation (40.10) we have

$$I_{n,n} = 2nI_{n-1,n-1}.$$

Repeating this operation n times and noting that

$$I_{0,0} = \int_{-\infty}^{\infty} e^{-x^2}\,\mathrm{d}x = \sqrt{\pi},$$

we find that

$$I_{n,n} = 2^n n! \sqrt{\pi}. \tag{40.12}$$

Combining equations (40.9) and (40.12) we have finally

$$I_{m,n} = 2^n n! \sqrt{\pi}\,\delta_{mn}. \tag{40.13}$$

The evaluation of more complicated integrals can be effected by combining this result with the recurrence formulae we have already established for the Hermite functions. For instance, it follows from equation (40.5) that

$$\int_{-\infty}^{\infty} x\Psi_m(x)\Psi_n(x)\,\mathrm{d}x = nI_{m,n-1}+\tfrac{1}{2}I_{m,n+1},$$

showing that

$$\int_{-\infty}^{\infty} x\Psi_m(x)\Psi_n(x)\,\mathrm{d}x = 0 \quad \text{if} \quad m\neq n\pm 1 \tag{40.14}$$

and that

$$\int_{-\infty}^{\infty} x\Psi_n(x)\Psi_{n+1}(x)\,\mathrm{d}x = 2^n(n+1)!\sqrt{\pi}. \tag{40.15}$$

Similarly, making use of equation (40.4) and equations (40.13–15) we can show that

$$\int_{-\infty}^{\infty} \Psi_m(x)\Psi_n'(x)\,\mathrm{d}x = \begin{cases} 0 & \text{if } m\neq n\pm 1 \\ 2^{n-1}n!\sqrt{\pi} & \text{if } m=n-1. \\ -2^n(n+1)!\sqrt{\pi} & \text{if } m=n+1 \end{cases}$$

§41. The occurrence of Hermite functions in wave mechanics

The Hermite functions which we have discussed in the last section occur in the wave mechanical treatment of the harmonic oscillator (see Mott and Sneddon, 1963). Although this is a very simple mechanical system, the analysis of its properties is of great importance because of its application to the quantum theory of radiation.

The Schrödinger equation corresponding to a harmonic oscillator of point mass m with vibrational frequency ν is

$$\frac{\mathrm{d}^2\psi}{\mathrm{d}x^2}+\frac{8\pi^2 m}{h^2}(W-2\pi^2 m\nu^2x^2)\psi = 0, \tag{41.1}$$

where W is the total energy of the oscillator and h is Planck's constant. The problem is to determine the wave functions ψ

which have the property that

(i) $\psi \to 0$ as $|x| \to \infty$;

(ii) $\int_{-\infty}^{\infty} |\psi|^2 \, dx = 1$.

If we let

$$x = \frac{1}{2\pi} \sqrt{\frac{h}{m\nu}} \, \xi,$$

then the equation (41.1) becomes

$$\frac{d^2\psi}{d\xi^2} + \left(\frac{2W}{h\nu} - \xi^2\right)\psi = 0 \tag{41.2}$$

and the conditions (i) and (ii) become

(i′) $\psi \to 0$ as $|\xi| \to \infty$;

(ii′) $\int_{-\infty}^{\infty} |\psi|^2 \, d\xi = 2\pi \sqrt{\frac{m\nu}{h}}$.

The argument given at the beginning of §40 shows that equation (41.2) possesses solutions ψ which satisfy the condition (ii′) if and only if the parameter $(2W/h\nu)$ which occurs in the equation takes one of the values $1+2n$, where n is a positive integer. In other words solutions of this type, which are known by the probability interpretation of the wave function ψ to correspond to stationary states of the system can exist if and only if

$$W = h\nu(n + \tfrac{1}{2}), \tag{41.3}$$

where n is a positive integer. When this is the case the form of the wave function ψ is known from §40 to be

$$\psi = C\Psi_n(\xi), \tag{41.4}$$

where C is a constant. Applying condition (ii′) and equation (40.12), we see that

$$C = \left(\frac{4\pi m\nu}{h}\right)^{\frac{1}{4}} \frac{1}{2^{\frac{1}{2}n}(n!)^{\frac{1}{2}}}. \tag{41.5}$$

Thus the wave function corresponding to an admissible energy $(n + \frac{1}{2})h\nu$ is

$$\psi_n = \left(\frac{4\pi m\nu}{h}\right)^{\frac{1}{4}} \frac{\Psi_n(\xi)}{2^{\frac{1}{2}n}(n!)^{\frac{1}{2}}}, \qquad \xi = 2\pi\sqrt{\frac{m\nu}{h}}\, x. \tag{41.6}$$

In quantum theory the matrix elements $(n\,|x|\,p)$ defined by the equation

$$(n\,|x|\,p)=\int_{-\infty}^{\infty} x\psi_n(x)\psi_p(x)\,\mathrm{d}x$$

are of considerable importance in the case of the harmonic oscillator. In terms of the variable ξ we have

$$(n\,|x|\,p)=\frac{h}{4\pi^2 m\nu}\int_{-\infty}^{\infty} \xi\psi_n(\xi)\psi_p(\xi)\,\mathrm{d}\xi,$$

so that substituting from equation (41.6) we have

$$(n\,|x|\,p)=\frac{1}{\pi}\sqrt{\frac{h}{4\pi m\nu}}\int_{-\infty}^{\infty} \xi\Psi_n(\xi)\Psi_p(\xi)\,\mathrm{d}\xi \div \{2^{\frac{1}{2}n+\frac{1}{2}p}(n!)^{\frac{1}{2}}(p!)^{\frac{1}{2}}\}.$$

It follows then from equations (40.14) and (40.15) that

$$(n\,|x|\,p)=0 \quad \text{if} \quad p\neq n\pm 1 \tag{41.7}$$

and that

$$(n\,|x|\,n+1)=\left\{\frac{(n+1)h}{8\pi^2 m\nu}\right\}^{\frac{1}{2}}, \tag{41.8}$$

$$(n\,|x|\,n-1)=\left\{\frac{nh}{8\pi^2 m\nu}\right\}^{\frac{1}{2}}. \tag{41.9}$$

§42. The Laguerre polynomials

The **Laguerre polynomials** $L_n(x)$ are defined for n a positive integer and x a positive real number by the equation

$$\exp\left(-\frac{xt}{1-t}\right)=(1-t)\sum_{n=0}^{\infty}\frac{L_n(x)}{n!}t^n. \tag{42.1}$$

Expanding the exponential function we see that

$$(1-t)^{-1}\exp\left(-\frac{xt}{1-t}\right)=\sum_{r=0}^{\infty}\frac{(-1)^r x^r t^r}{r!(1-t)^{r+1}}$$

$$=\sum_{r=0}^{\infty}\sum_{s=0}^{\infty}\frac{(-1)^r(r+1)s}{r!s!}x^r t^{r+s}.$$

The coefficient of t^n in this expansion is

$$\sum_{r=0}^{n} \frac{(-1)^r (r+1)_{n-1}}{r!(n-r)!} x^r.$$

Using the relations

$$(r+1)_{n-r} = \frac{n!}{r!}, \qquad \frac{(-1)^r}{(n-r)!} = \frac{(-n)r}{n!},$$

we see that this sum can be written in the form

$$\sum_{r=0}^{n} \frac{(-n)_r}{(r!)^2} x^r.$$

Identifying the coefficient of t^n with $L_n(x)/n!$ and adopting the notation of §11, we see that

$$L_n(x) = n!\,{}_1F_1(-n; 1; x). \tag{42.2}$$

It should be observed that $L_n(x)$ is a polynomial of degree n in x, and that the coefficient of x^n is $(-1)^n$.

A useful formula for the polynomial $L_n(x)$ can be obtained by finding a new representation for the confluent hypergeometric function on the right-hand side of equation (42.2). By Leibnitz's theorem for the nth derivative of a product of two functions we have

$$e^x D^n(x^n e^{-x}) = e^x \sum_{r=0}^{n} (-1)^r \frac{(-n)_r}{r!} (D^{n-r} x^n)(D^r e^{-x}),$$

where D denotes the operator d/dx. Using the relations

$$e^x D^r(e^{-x}) = (-1)^r, \qquad D^{n-r} x^n = n! x^r / r!,$$

we see that

$$e^x D^n(x^n e^{-x}) = n! \sum_{r=0}^{n} \frac{(-n)_r}{(r!)^2} (x)^r. \tag{42.3}$$

It follows immediately from equation (42.2) that

$$L_n(x) = e^x \frac{\mathrm{d}^n}{\mathrm{d}x^n} (x^n e^{-x}). \tag{42.4}$$

The first five Laguerre polynomials can be calculated easily from this equation; we find that

$$\begin{aligned}
L_0(x)&=1,\\
L_1(x)&=1-x,\\
L_2(x)&=2-4x+x^2,\\
L_3(x)&=6-18x+9x^2-x^3,\\
L_4(x)&=24-96x+72x^2-16x^3+x^4.
\end{aligned}$$

Equations (42.4) can be used to show that the functions

$$\phi_n(x)=\frac{1}{n!}e^{-\frac{1}{2}x}L_n(x) \tag{42.5}$$

form an orthonormal system. From (42.4) we have as a result of m integrations by parts

$$\begin{aligned}
\int_0^\infty e^{-x}x^m L_n(x)\,\mathrm{d}x &= \int_0^\infty x^m \frac{\mathrm{d}^n}{\mathrm{d}x^n}(x^n e^{-x})\,\mathrm{d}x\\
&=(-1)^m m!\int_0^\infty \frac{\mathrm{d}^{n-m}}{\mathrm{d}x^{n-m}}(x^n e^{-x})\,\mathrm{d}x
\end{aligned}$$

and this is zero if $n>m$. Since $L_m(x)$ is a polynomial of degree m in x it follows that

$$\int_0^\infty e^{-x}L_m(x)L_n(x)\,\mathrm{d}x=0 \quad \text{if} \quad m\neq n. \tag{42.6}$$

Since the term of degree n in $L_n(x)$ is $(-1)^n x^n$ it follows that, when $m=n$,

$$\begin{aligned}
\int_0^\infty e^{-x}\{L_n(x)\}^2\,\mathrm{d}x &= (-1)^n\int_0^\infty e^{-x}x^n L_n(x)\,\mathrm{d}x\\
&=\int_0^\infty n!x^n e^{-x}\,\mathrm{d}x\\
&=(n!)^2.
\end{aligned}$$

Combining this result with equation (42.6) we find that

$$\int_0^\infty \phi_m(x)\phi_n(x)\,\mathrm{d}x=\delta_{mn}, \tag{42.7}$$

showing that the ϕ's form an orthonormal set.

Recurrence formulae for the Laguerre polynomials may be derived directly from the definition (42.1). Differentiating both sides of this equation with respect to t we obtain the identity

$$-\frac{x}{(1-t)^2}\exp\left(-\frac{xt}{1-t}\right)=(1-t)\sum_{n=1}^{\infty}\frac{L_n(x)t^{n-1}}{(n-1)!}-\sum_{n=0}^{\infty}\frac{L_n(x)t^n}{n!},$$

which may be written in the form

$$x\sum_{n=0}^{\infty}\frac{L_n(x)t^n}{n!}+(1-t)^2\sum_{n=0}^{\infty}\frac{L_n(x)t^{n-1}}{(n-1)!}-(1-t)\sum_{n=0}^{\infty}\frac{L_n(x)t^n}{n!}=0.$$

Equating to zero the coefficient of t^n in the expansion on the left we obtain the recurrence relation

$$L_{n+1}(x)+(x-2n-1)L_n(x)+n^2L_{n-1}(x)=0. \tag{42.8}$$

Similarly if we differentiate both sides of (42.1) with respect to x we obtain the identity

$$t\sum_{n=0}^{\infty}\frac{L_n(x)}{n!}t^n+(1-t)\sum_{n=0}^{\infty}\frac{L_n'(x)}{n!}t^n=0,$$

which yields the recurrence relation

$$L_n'(x)-nL_{n-1}'(x)+nL_{n-1}(x)=0. \tag{42.9}$$

Differentiating equation (42.8) twice with respect to x and replacing n by $n+1$, we find that

$$L_{n+2}''(x)+(x-2n-3)L_{n+1}''(x)+(n+1)^2L_n''(x)+2L_{n+1}'(x)=0. \tag{42.10}$$

Now from (42.9)

$$L_{n+1}'(x)=(n+1)\{L_n'(x)-L_n(x)\},$$

and hence

$$L_{n+1}''(x)=(n+1)\{L_n''(x)-L_n'(x)\}.$$

A similar expression for $L_{n+2}''(x)$ in terms of $L_n(x)$ and its derivatives can be readily obtained. Substituting these values of $L_{n+1}''(x)$ and $L_{n+1}'(x)$ in equation (42.10), we find that

$$xL_n''(x)+(1-x)L_n'(x)+nL_n(x)=0. \tag{42.11}$$

§43. Laguerre's differential equation

Equation (42.8) shows that $y = AL_n(x)$ is a solution of **Laguerre's differential equation**

$$x\frac{d^2y}{dx^2} + (1-x)\frac{dy}{dx} + \nu y = 0 \tag{43.1}$$

in the case in which ν is a positive integer n. If we put $\gamma = 1$, $\alpha = -\nu$ in equation (11.2), we see that it takes the form (43.1), so that it follows from equation (11.4) that one solution of equation (43.1) is

$$y_1(x) = {}_1F_1(-\nu; 1; x). \tag{43.2}$$

Similarly we see from equation (11.8) that the second solution is

$$y_2(x) = y_1(x)\log x + \sum_{r=1}^{\infty} c_r x^r, \tag{43.3}$$

where the coefficients c_r are defined by equations of the type (11.9). The general solution of Laguerre's differential equation may therefore be written in the form

$$y = Ay_1(x) + By_2(x),$$

where A and B are constants and $y_1(x)$, $y_2(x)$ are defined by equations (43.2) and (43.3) respectively.

If we are interested only in solutions which remain finite at $x = 0$ it is obvious from equation (43.3) that we must take the constant B to be zero. Further, if a_r is the coefficient of x^r in the series expansion for $y_1(x)$, it is easily shown that $a_{r+1}/a_r \sim r^{-1}$. As the result of a discussion similar to that advanced in §39, it follows that if ν is not an integer

$$y_1(x) \sim e^x \quad \text{as} \quad x \to \infty.$$

If, therefore, we are looking for solutions which increase less rapidly than this we must take ν to a positive integer, in which case $y_1(x)$ reduces to a polynomial.

The equation (43.1) possesses a solution which increases less rapidly than e^x as $x \to \infty$ if and only if the parameter ν occurring in it is a positive integer, n say. If it is also required that the solution shall remain finite at $x = 0$, the solution is of the form

$$y = AL_n(x), \tag{43.4}$$

where A is a constant and $L_n(x)$ is the Laguerre polynomial of degree n.

By making use of the methods described in §3, we can, of course, derive contour integral solutions of Laguerre's equation. The basic results are contained in Problem 5.9.

§44. The associated Laguerre polynomials and functions

If we differentiate Laguerre's differential equation m times with respect to x, we find that it becomes

$$x\frac{d^{m+2}y}{dx^{m+2}}+(m+1-x)\frac{d^{m+1}y}{dx^{m+1}}+(n-m)\frac{d^m y}{dx^m}=0.$$

which shows that $L_n^m(x)$ defined by

$$L_n^m(x)=\frac{d^m}{dx^m}L_n(x), \qquad n\geqslant m, \tag{44.1}$$

is a solution of the differential equation

$$x\frac{d^2y}{dx^2}+(m+1-x)\frac{dy}{dx}+(n-m)y=0. \tag{44.2}$$

The polynomial $L_n^m(x)$ defined by equation (44.1) is called the **associated Laguerre polynomial.** It follows from equation (42.2) that it may be presented as a series by the equation

$$L_n^m(x)=\frac{(-1)^m(n!)^2}{m!(n-m)!}\,{}_1F_1(-n+m;m+1;x), \quad n\geqslant m. \tag{44.3}$$

Similarly equation (42.4) leads to the formula

$$L_n^m(x)=\frac{d^m}{dx^m}\left\{e^x\frac{d^n}{dx^n}(x^n e^{-x})\right\}. \tag{44.4}$$

The simplest associated Laguerre polynomials are:

$L_1^1(x)=-1,$

$L_2^1(x)=-4+2x, \qquad L_2^2(x)=2,$

$L_3^1(x)=-18+18x-3x^2, \qquad L_3^2(x)=18-6x, \qquad L_3^3(x)=-6,$

$L_4^1(x)=-96+144x-48x^2+4x^3, \qquad L_4^2(x)=144-96x+12x^2$

$L_4^3(x)=-96+24x, \qquad L_4^4(x)=24.$

The definition (44.1) for the associated Laguerre polynomial is the one usually taken in applied mathematics. In pure mathematics the function

$$L_n^{(m)}(x)=\frac{(m+n)!}{m!n!}\,{}_1F_1(-n;m+1;x) \tag{44.5}$$

which is a solution of the differential equation

$$x\frac{d^2y}{dx^2}+(m+1-x)\frac{dx}{dy}+ny=0$$

is often taken as the definition of the associated Laguerre polynomial (for instance, in Copson (1935)), so that care must be taken in reading the literature to ensure that the particular convention being followed is understood.

It is readily shown from equation (42.1), which defines the generating function for Laguerre polynomials, that the associated Laguerre polynomials may be defined by the equation

$$(-1)^m t^m \exp\left(-\frac{xt}{1-t}\right)=(1-t)^{m+1}\sum_{n=m}^{\infty}\frac{L_n^m(x)}{n!}t^n. \tag{44.6}$$

This identity can then be used to derive recurrence relations for the associated Laguerre polynomials similar to those of equations (42.8) and (42.9) (cf. Problem 6(ii), (iii) below).

The **Laguerre functions** $R_{nl}(x)$ are defined by the equation

$$R_{nl}(x)=e^{-\frac{1}{2}x}x^l L_{n+1}^{2l+1}(x), \qquad n\geqslant l+1. \tag{44.7}$$

If in equation (44.2) we replace m by $2l+1$, n by $n+1$ and y by $e^{\frac{1}{2}x}x^{-l}R$, we see that the function $R_{nl}(x)$ is a solution of the ordinary linear differential equation

$$\frac{d^2R}{dx^2}+\frac{2}{x}\frac{dR}{dx}-\left\{\frac{1}{4}-\frac{n}{x}+\frac{l(l+1)}{x^2}\right\}R=0. \tag{44.8}$$

In most physical problems in which this equation arises it is known that l is an integer. By reasoning similar to that outlined in §43 it can readily be shown that the equation (44.8) possesses a solution which is finite at $x=0$ and tends to zero as $x\to\infty$ if, and only if, the parameter n which occurs in it assumes integral values. When this does occur the solution is $AR_{nl}(x)$, where the function $R_{nl}(x)$ is given by equation (44.7) and A is an arbitrary constant.

We shall now evaluate the integral

$$I_{nl} = \int_0^\infty x^2\{R_{nl}(x)\}^2\,\mathrm{d}x$$

which arises in wave mechanics. From equation (44.6) we have the identity

$$\sum_{n=l+1}^{\infty}\sum_{n'=l+1}^{\infty}\frac{L_{n+l}^{2l+1}(x)L_{n'+l}^{2l+1}(x)}{(n+l)!(n'+l)!}t^{n+l}\tau^{n'+l}=\frac{\exp\left\{-\dfrac{xt}{1-t}-\dfrac{x\tau}{1-\tau}\right\}}{(1-t)^{2l+2}(1-\tau)^{2l+2}}t^{2l+1}\tau^{2l+1},$$

from which it follows that

$$\sum_{n=l+1}^{\infty}\sum_{n'=l+1}^{\infty}\frac{t^{n+l}\tau^{n'+l}}{(n+l)!(n'+l)!}\int_0^\infty e^{-x}x^{2l+2}L_{n+l}^{2l+1}(x)L_{n'+l}^{2l+1}(x)\,\mathrm{d}x$$
$$=\frac{(t\tau)^{2l+1}}{(1-t)^{2l+2}(1-\tau)^{2l+2}}\int_0^\infty x^{2l+2}\exp\left\{-x-\frac{xt}{1-t}-\frac{x\tau}{1-\tau}\right\}\mathrm{d}x.$$

This last integration is elementary and gives for the expression on the right

$$\frac{(2l+2)!(t\tau)^{2l+1}(1-t)(1-\tau)}{(1-t\tau)^{2l+3}}$$

and by means of the binomial theorem we may expand this function in the form

$$(1-t-\tau-t\tau)\sum_{r=0}^{\infty}\frac{(2l+r+2)!}{r!}(t\tau)^{2l+r+1}.$$

Now the coefficient of $(t\tau)^{n+1}$ in this expansion is

$$\frac{2n\{(n+l)!\}}{(n-l-1)!}$$

and this is equal to $I_{nl}/\{(n+l)!\}^2$. Hence

$$\int_0^\infty x^2\{R_{nl}(x)\}^2\,\mathrm{d}x=\frac{2n\{(n+l)!\}^3}{(n-l-1)!}. \tag{44.9}$$

§45. The wave functions for the hydrogen atom

We shall conclude this chapter by discussing the motion of a single electron of mass m and charge $-e$ in the Coulomb field of force with potential

$$V(r) = -\frac{Ze^2}{r},$$

due to a nucleus of charge Ze. In a first approximation we may treat the mass of the nucleus as infinite and in this case the wave function ψ of the system is governed by the Schrödinger equation

$$\nabla^2\psi + \frac{8\pi^2 m}{h^2}\left(W + \frac{Ze^2}{r}\right)\psi = 0 \tag{45.1}$$

and the conditions:

(i) $\psi(r, \theta, \phi + 2\pi) = \psi(r, \theta, \phi)$ for all r, θ, ϕ;
(ii) ψ must be bounded in the range $0 \leq \theta \leq \pi$ for all r, ϕ;
(iii) $\psi \to 0$ as $r \to \infty$;
(iv) ψ remains finite as $r \to 0$;
(v) ψ is normalised to unity, i.e. $\int |\psi|^2 \, d\tau = 1$, where the integral is taken throughout the whole of space.

W is the total energy of the system.

The equation (45.1) may be solved by setting

$$\psi = R(r)\Theta(\theta)\Phi(\phi), \tag{45.2}$$

where, by the method of separation of variables, we have

$$\frac{d^2\Phi}{d\phi^2} + u^2\Phi = 0, \tag{45.3}$$

$$\frac{1}{\sin\theta}\frac{d}{d\theta}\left(\sin\theta\frac{d\Theta}{d\theta}\right) + \left\{l(l+1) - \frac{u^2}{\sin^2\theta}\right\}\Theta = 0, \tag{45.4}$$

$$\frac{1}{r^2}\frac{d}{dr}\left(r^2\frac{dR}{dr}\right) + \left\{\frac{8\pi^2 m}{h^2}\left(W + \frac{Ze^2}{r}\right) - \frac{l(l+1)}{r^2}\right\}R = 0. \tag{45.5}$$

To satisfy condition (i) we must choose as a solution of (45.3) a function $\Phi(\phi)$ such that $\Phi(\phi + 2\pi) = \Phi(\phi)$ for all ϕ. Thus u occurring in equation (45.3) must be an integer and a convenient solution will be

$$\Theta = Ae^{iu\phi}, \tag{45.6}$$

where A is an arbitrary constant. Equation (45.5) is the well-known equation of which the solution is the associated Legendre polynomial $P_l^u(\cos\theta)$. If l is integral and $l \geq |u|$, then $P_l^u(\cos\theta)$ is the only solution which is bounded in the range $0 \leq \theta \leq \pi$, and is therefore the only one leading to a wave function ψ which satisfies condition (ii) above; if l is not an integer no bounded solution exists.

To solve equation (45.5) we write

$$\alpha^2 = -\frac{8\pi^2 mW}{h^2}, \qquad \nu = \frac{4\pi^2 mZe^2}{h^2\alpha} \tag{45.7}$$

and change the independent variable from r to x where $x = 2\alpha r$. We then find that

$$\frac{d^2R}{dx^2} + \frac{2}{x}\frac{dR}{dx} - \left\{\frac{1}{4} - \frac{\nu}{x} + \frac{l(l+1)}{x^2}\right\}R = 0 \tag{45.8}$$

where, in order that conditions (iii) and (iv) be satisfied, R must be such that $R \to 0$ as $x \to \infty$ and as $x \to 0$. From the arguments of §44 it follows that this is possible only if l is a positive integer and only if ν is an integer, n say, which is greater than $l+1$. When this is so we may write the solutions of equation (45.8) in the form (44.7), so that the solution of (45.5) is proportional to $R_{nl}(2\alpha r)$. Now by the second of the equations (45.7), we have

$$W = -\frac{2\pi^2 Z^2 e^4 m}{h^2 n^2} \tag{45.9}$$

for the possible values of the total energy W. The wave functions corresponding to the value of energy given by the integer n are

$$\psi_{nlu}(r, \theta, \phi) = C_{nlu} R_{nl}(2\alpha r) P_l^u(\cos\theta) e^{iu\phi}, \tag{45.10}$$

where C_{nlu} is a constant determined by the condition (v). In polar coordinates $d\tau = r^2 \sin\theta \, dr \, d\theta \, d\phi$, so that this condition gives

$$1 = C_{nlu}^2 \int_0^{2\pi} d\phi \int_0^{\pi} \sin\theta \{P_l^u(\cos\theta)\}^2 \, d\theta \int_0^{\infty} r^2 R_{nl}^2(2\alpha r) \, dr,$$

so that it follows from equations (21.20) and (44.9) that

$$C_{nlu} = (2\alpha)^{\frac{3}{2}} \left\{\frac{(-1)^u (l-u)!(n-l-1)!(2l+1)}{8\pi u(l+u)!\{(n+l)!\}^3}\right\}^{\frac{1}{2}}. \tag{45.11}$$

Now from equations (45.7) and (45.9) we find that $\alpha = 4\pi^2 Ze^2 m/(h^2 n)$, so that if we introduce the Bohr radius a by means of the equation

$$a = \frac{h^2}{4\pi^2 me^2}, \tag{45.12}$$

we find that $\alpha = Z/an$. Introducing this result into (45.11) and substituting the value obtained for C_{nlu} into equation (45.10), we find that

$$\psi_{nlu}(r,\theta,\phi)\left\{\left(\frac{2Z}{an}\right)^3\frac{(l-u)!(n-l-1)!(2l+1)}{(l+u)!\{(n+l)!\}^3 8\pi n}\right\}^{\frac{1}{2}} R_{nl}(2\alpha r)P_l^n(\cos\theta)e^{iu\phi}, \tag{45.13}$$

with $u \leqslant l \leqslant n-1$, are the wave-functions corresponding to the energy (45.9) of the hydrogen atom.

If we write $W_0 = -2\pi^2 Z^2 e^4 m/h^2$, then corresponding to the energy W_0 we have the wave function

$$\psi_{100}(r,\theta,\phi)=\frac{1}{\sqrt{\pi}}\left(\frac{Z}{a}\right)^{\frac{3}{2}}e^{-\rho}, \qquad \rho = Zr/a,$$

and to the energy level $W_0/4$ we have the three wave functions

$$\psi_{200}(r,\theta,\phi)=\frac{1}{4\sqrt{(2\pi)}}\left(\frac{Z}{a}\right)^{\frac{3}{2}}(2-\rho)e^{-\frac{1}{2}\rho},$$

$$\psi_{210}(r,\theta,\psi)=\frac{1}{4\sqrt{(2\pi)}}\left(\frac{Z}{a}\right)^{\frac{3}{2}}\rho e^{-\frac{1}{2}\rho}\cos\theta,$$

$$\psi_{211}(r,\theta,\phi)=\frac{1}{8\sqrt{(\pi)}}\left(\frac{Z}{a}\right)^{\frac{3}{2}}\rho e^{-\frac{1}{2}\rho}\sin\theta\, e^{\pm l\phi}.$$

From the last of these functions we can construct two functions

$$\frac{1}{4\sqrt{2\pi}}\left(\frac{Z}{a}\right)^{\frac{3}{2}}\rho e^{-\frac{1}{2}\rho}\sin\theta \begin{matrix}\sin\\ \\ \cos\end{matrix}\phi.$$

Similarly, corresponding to the energy $W_0/9$, we have the six wave functions

$$\psi_{300}(r,\theta,\phi)=\frac{1}{81\sqrt{3\pi}}\left(\frac{Z}{a}\right)^{\frac{3}{2}}(27-18\rho+2\rho^2)e^{-\frac{1}{3}\rho},$$

$$\psi_{310}(r,\theta,\phi)=\frac{\sqrt{2}}{81\sqrt{\pi}}\left(\frac{Z}{a}\right)^{\frac{3}{2}}(6-\rho)\rho e^{-\frac{1}{3}\rho}\cos\theta,$$

$$\psi_{311}(r,\theta,\phi)=\frac{1}{81\sqrt{\pi}}\left(\frac{Z}{a}\right)^{\frac{3}{2}}(6-\rho)\rho e^{-\frac{1}{3}\rho}\sin\theta\, e^{\pm i\phi},$$

$$\psi_{320}(r,\theta,\phi)=\frac{1}{81\sqrt{6\pi}}\left(\frac{Z}{a}\right)^{\frac{3}{2}}\rho^2 e^{-\frac{1}{3}\rho}(3\cos^2\theta-1),$$

$$\psi_{321}(r,\theta,\phi)=\frac{1}{81\sqrt{\pi}}\left(\frac{Z}{a}\right)^{\frac{3}{2}}\rho^2 e^{-\frac{1}{3}\rho}\sin\theta\cos\theta\, e^{\pm i\phi},$$

$$\psi_{322}(r,\theta,\phi)=\frac{1}{162\sqrt{\pi}}\left(\frac{Z}{a}\right)^{\frac{3}{2}}\rho^2 e^{-\frac{1}{3}\rho}\sin^2\theta\, e^{\pm 2i\phi}.$$

In wave mechanics $|\psi(r, \theta, \phi)|^2 \, d\tau$ represents the probability that the electron whose wave-function is $\psi(r, \theta, \phi)$ is to be found in a small volume $d\tau$ centred at the point whose polar coordinates are (r, θ, ϕ). To make the total probability unity we must have

$$\int |\psi(r, \theta, \phi)|^2 \, d\tau = 1,$$

where the integral is taken throughout the whole space. Since, in polar coordinates, $d\tau = r^2 \sin\theta \, dr \, d\theta \, d\phi$, it follows that the probability that the electron is at a distance between $r - \frac{1}{2} dr$ and $r + \frac{1}{2} dr$ from the nucleus is $\phi(r) \, dr$, where

$$\phi(r) = r^2 \int_0^{2\pi} d\phi \int_0^{\pi} \sin\theta \, d\theta \, . \, |\psi(r, \theta, \phi)|^2.$$

Hence for an electron in the state defined by quantum numbers n, l, u (equation (45.13) above) we have

$$\phi(r) \equiv \left(\frac{2Z}{an}\right)^3 \frac{(n-l-1)!}{2n\{(n+l)!\}^3} r^2 \{R_{nl}(2\alpha r)\}^2.$$

The mean value of any function f which depends on r alone is given by

$$\bar{f} = \int_0^\infty f(r)\phi(r) \, dr,$$

that is, by the formula

$$\bar{f} = \left(\frac{2Z}{an}\right)^3 \frac{(n-l-1)!}{2n\{(n+l)!\}^3} \int_0^\infty r^2 f(r) \{R_n(2\alpha r)\}^2 \, dr.$$

For examples of the use of this formula see Problem 5.21.

Problems V

5.1 Prove that if $m < n$,

$$\frac{d^m}{dx^m}\{H_n(x)\} = \frac{2^m n!}{(n-m)!} H_{n-m}(x).$$

5.2 Show that

$$\text{(i)} \; 2^{n+1} n! \sum_{r=0}^{n} \frac{H_r(x) H_r(y)}{2^r r!} = \frac{H_{n+1}(x) H_n(y) - H_n(x) H_{n+1}(y)}{x - y},$$

(ii) $$2^{n+1}n!\sum_{r=0}^{n}\frac{H_r(x_k)H_r(x)}{2^r r!}=H_{n+1}(x_k)\frac{H_n(x)}{x_k-x},$$

(iii) $$\int_{-\infty}^{\infty}\frac{H_n(x)}{x-x_k}e^{-x^2}\,dx=\frac{2^n(n-1)!\pi^{\frac{1}{2}}}{H_{n-1}(x_k)}.$$

Deduce the value of the weights $\{w_k\}_{k=1}^{n}$ in the Gaussian integration formula

$$\int_{-\infty}^{\infty}e^{-x^2}f(x)\,dx=\sum_{k=1}^{n}w_k f(x_k)+R_n(f).$$

5.3 Show that Hermite's equation

$$w''(z)-2zw'(z)+2\nu w(z)=0$$

has the contour integral solution

$$w(z)=A\int_C \exp\left(-\tfrac{1}{2}s^2+isz\right)s^{2\nu-1}\,ds,$$

where A is an arbitrary constant and the contour C is such that

$$\Delta_C[s^{2\nu}\exp\left(-\tfrac{1}{2}s^2+isz\right)]=0.$$

Show also that Hermite's equation has a solution

$$w(z)=Ae^{z^2}\int_C\frac{e^{-s^2}\,ds}{(s-z)^{\nu+1}},$$

where, again, A is an arbitrary constant but now the condition on C is that

$$\Delta_C[(s-z)^{-\nu-2}e^{-s^2}]=0.$$

5.4 Prove that if

$$K(x,y,t)=\sum_{n=0}^{\infty}\psi_n(x)\psi_n(y)t^n,$$

where the $\psi_n(x)$ are the orthonormal set of Hermite functions defined by the relation

$$\psi_n(x)=2^{-\frac{1}{2}n}(n!)^{-\frac{1}{2}}\pi^{-\frac{1}{4}}\Psi_n(x),$$

then

$$\int_{-\infty}^{\infty}\exp\left(-\tfrac{1}{2}x^2+2\lambda x-\lambda^2\right)K(x,y,t)\,dx=\exp\left(-\tfrac{1}{2}y^2+2\lambda ty-\lambda^2t^2\right).$$

Assuming that $K(x, y, t)$ is of the form

$$A \exp(Bx^2+Cxy+Dy^2),$$

where A, B, C and D are functions of t, prove that

$$K(x, y, t)=\frac{1}{\sqrt{\{\pi(1-t)^2\}}}\exp\left\{\frac{4xyt-(x^2+y^2)(1+t^2)}{2(1-t)^2}\right\}.$$

5.5 Show that

$$\sum_{n=0}^{\infty}\{H_n(x)\}^2\frac{t^n}{n!}=\frac{1}{\sqrt{(1-4t^2)}}\exp\left\{\frac{4x^2t}{1+t}\right\}.$$

5.6 The Schrödinger equation for the three-dimensional oscillator in cylindrical coordinates (ρ, ϕ, z) is

$$\nabla^2\psi+\frac{8\pi^2m}{h^2}\{W-2\pi^2m(\nu^2\rho^2+\omega^2z^2)\}\psi=0.$$

Show that the solutions ψ of this equation such that $\psi\to 0$ as ρ and $z\to\infty$ and ψ is finite at the origin are of the form

$$C\exp\{iu-\tfrac{1}{2}\alpha^2\rho^2-\tfrac{1}{2}\beta^2z^2\}f_{l\,|u|}(\xi)H_n(z),$$

where C is a constant, l, n are positive integers and u is an integer, $\alpha=2\pi\sqrt{(m\nu/h)}$, $\beta=2\pi\sqrt{(m\omega/h)}$, and $f_{l\,|u|}(\xi)$ is a polynomial of degree $2l$ in ξ which satisfies the equation

$$\frac{d^2f}{d\xi^2}-\left(2\xi-\frac{1}{\xi}\right)\frac{df}{d\xi}+\left(2\,|u|+2l-\frac{u^2}{\xi^2}\right)f=0.$$

Show also that the corresponding values of W are given by the equation

$$W=(2l+|u|+1)h\nu+(n+\tfrac{1}{2})h\omega.$$

5.7 Prove that

$$L_n(2x)=n!\sum_{m=0}^{n}\frac{2^{n-m}(-1)^m}{m!(n-m)!}L_{n-m}(x).$$

5.8 Show that

(i) $$(n!)^2\sum_{r=0}^{n}\frac{L_r(x)L_r(y)}{(r!)^2}=\frac{L_{n+1}(x)L_n(y)-L_n(x)L_{n+1}(y)}{x-y},$$

(ii) $$(r!)^2\sum_{r=0}^{n}\frac{L_r(x_k)L_r(x)}{(r!)^2}=L_{n+1}(x_k)\frac{L_n(x)}{x_k-x},$$

(iii) $\displaystyle\int_0^\infty e^{-x}L_n(x)\frac{dx}{x-x_k}=\frac{\{(n-1)!\}^2}{L_{n-1}(x_k)}$,

where $\{x_k\}_{k=1}$ are the zeros of $L_n(x)$.

Deduce the value of the weights $\{w_k\}_{k=1}$ in the Gaussian integration formula

$$\int_0^\infty e^{-x}f(x)\,dx=\sum_{k=1}^{n} w_k f(x_k)+R_n(f).$$

5.9 Prove that

$$w(z)=A\int_C p^{-\nu-1}(p-1)^\nu e^{pz}\,dp,$$

in which A is an arbitrary constant, is a solution of Laguerre's differential equation

$$zw''(z)+(1-z)w'(z)+\nu w(z)=0,$$

provided that the contour C is such that

$$\Delta_C[p^{-\nu}(p-1)^{\nu+1}e^{pz}]=0.$$

Show also that Laguerre's equation has a solution

$$w(z)=Ae^z\int_C\frac{e^{-s}\,ds}{(s-z)^{\nu+1}},$$

where, as before, A is an arbitrary constant but now the condition on C is that

$$\Delta_C[(s-z)^{-\nu-2}e^{-s}]=0.$$

5.10 Prove that

(i) $\dfrac{d}{dx}L_n^m(x)=L_n^{m+1}(x)$;

(ii) $L_{n+1}^m(x)+(x-2n-1)L_n^m(x)+mL_n^{m-1}(x)+n^2L_{n-1}^m(x)=0$;

(iii) $L_n^m(x)-nL_{n-1}^m(x)+nL_{n-1}^{m-1}(x)=0$.

5.11 Prove that if $m\leqslant n$,

$$\int_0^\infty J_m(2\sqrt{x})e^{-x/a}x^{n-\frac{1}{2}m}\,dx=(-1)^m\frac{(n-m)!}{n!}a^{n+1}e^{-a}L_n^m(a).$$

5.12 Prove that if $m\leqslant n$, $a>0$,

$$\int_0^\infty e^{-ax}x^mL_n^m(x)\,dx=\frac{(-1)^m(n!)^2(a-1)^{n-m}}{(n-m)!a^{n+1}},$$

and deduce that if $n \geq l+1$

$$\int_0^\infty x^{l+1} R_{nl}(x)\, dx = \frac{\{(n+l)!\}^2}{(n-l-1)!}(-1)^{n+l}2^{2l+2}.$$

5.13 Show that

$$\int_0^1 x^m(1-x)^p L_n^m(ax)\, dx = (-1)^{p+1}\frac{(n!)^2 p!}{\{(n+p)!\}} L_{n+p+1}^{m+p+1}(a).$$

5.14 Prove that the function

$$x^{\frac{1}{2}\gamma} e^{-\frac{1}{2}x} L_{\alpha+\beta}^{\alpha}(x)$$

is a solution of the differential equation

$$\frac{d^2y}{dx^2} + \frac{\alpha-\gamma+1}{x}\frac{dy}{dx} + \left\{-\frac{1}{4} + \frac{2\beta+\alpha+1}{2x} - \frac{\gamma(2\alpha-\gamma)}{4x^2}\right\} y = 0.$$

5.15 The potential energy for the nuclear motion of a diatomic molecule is closely approximated by the Morse function

$$V(r) = De^{-2ar} - 2De^{-ar}.$$

Show that the spherically symmetrical solutions of the Schrödinger equation with this potential are

$$\psi(r) = \frac{C_n}{r}\{\exp(-be^{-ar})\}(2br)^{k-n-\frac{1}{2}} L_{2k-n+1}^{2k-2j-1}(2be^{-ar}), \quad 0 \leq n \leq k-\tfrac{1}{2},$$

where $b = 2\pi(2mD)/(ah)$, C_n is a normalisation constant, and the corresponding values of the energy are

$$W = -D\left[1 - \frac{(n+\frac{1}{2})}{b} + \frac{(n+\frac{1}{2})^2}{4b^2}\right].$$

5.16 Prove that the normalisation constant C_n in Problem 5.15 has the value

$$\sqrt{\frac{2ba}{N_n}},$$

where

$$N_m = \{\Gamma(2b-n)\}^2 \sum_{r=0}^{n} \frac{\Gamma(2b-2n+r-1)}{r!}.$$

5.17 Show that in parabolic coordinates ξ, η, ϕ defined by the equations

$$x = (\xi\eta)^{\frac{1}{2}}\cos\phi, \qquad y = (\xi\eta)^{\frac{1}{2}}\sin\phi, \qquad z = \tfrac{1}{2}(\xi-\eta),$$

Schrödinger's equation for a hydrogen-like atom of nuclear charge Ze takes the form

$$\frac{\partial}{\partial \xi}\left(\xi\frac{\partial \psi}{\partial \xi}\right)+\frac{\partial}{\partial \eta}\left(\eta\frac{\partial \psi}{\partial \eta}\right)+\frac{1}{4}\left(\frac{1}{\xi}+\frac{1}{\eta}\right)\frac{\partial^2 \psi}{\partial \phi^2}$$
$$+\frac{2\pi^2 m}{h^2}\{W(\xi+\eta)+2Ze^2\}\psi=0.$$

Show that this equation has solutions of the form

$$\psi(\xi, \eta, \phi)=C\left(\frac{\xi\eta}{n^2a^2}\right)e^{iu\phi-(\xi+\eta)/2na}L_{k+u}^{u}\left(\frac{\xi}{na}\right)L_{l+u}^{u}\left(\frac{\eta}{na}\right),$$

where a is the Bohr radius, $n=k+l+a+1$ and $W=-2\pi^2me^4z^2/(n^2/h^2)$.

Determine the constant C so that ψ is normalised to unity.

5.18 In the theory of the rotation and vibration spectrum of a diatomic molecule there arises the problem of solving the Schrödinger equation with potential energy

$$V(r)=\frac{B}{r^2}-\frac{Ze^2}{r}.$$

Show that the energy levels are given by

$$W=-2\pi^2mZ^2e^4/(h^2\sigma^2),$$

where $\sigma=n+\frac{1}{2}+\sqrt{\{b+(l+\frac{1}{2})^2\}}$ with n, l integers and $b=8\pi^2mB/h^2$, and that the corresponding wave functions

$$C_{n\alpha}\left(\frac{2r}{\sigma a}\right)^{\frac{1}{2}\alpha-\frac{1}{2}}e^{-r/\sigma a}L_{n+\alpha}^{\alpha}\left(\frac{2r}{\sigma a}\right)P_l^u(\cos\theta)e^{iu\phi},$$

where a is the Bohr radius, $\alpha=2\sqrt{\{b+(l+\frac{1}{2})^2\}}$ and $C_{n\alpha}$ is a normalisation factor.

5.19 Show that the constant $C_{n\alpha}$ occurring in Problem 5.18 has the value

$$\left\{\frac{2}{\sigma a\Gamma(n+\alpha+1)}\right\}^{\frac{3}{2}}\left\{\frac{(2l+1)n!(l-u)!}{4\pi(2n+\alpha+1)(l+u)!}\right\}^{\frac{1}{2}}.$$

5.20 If

$$I(l; n; s)=\int_0^\infty x^{s+2}\{R_{nl}(x)\}^2\,dx,$$

prove that $I(l;n;s) \div \{(n+l)!\}^2$ is the coefficient of $(t\tau)^{n+1}$ in the expansion of

$$(1-t)^{s+1}(1-\tau)^{s+1}\sum_r \frac{(2l+s+r+2)!}{r!}(t\tau)^{2l+r+1}.$$

Deduce that

$$I(l;n;-1)=\frac{\{(n+l)!\}^3}{(n-l-1)!},$$

$$I(l;n;1)=\frac{\{(n+l)!\}^3}{(n-l-1)!}\{6n^2-2l(l+1)\}.$$

5.21 Show that in a hydrogen-like atom of nuclear charge Ze the average distance of the electron from the nucleus, in the state described by quantum numbers l, n, is

$$\frac{n^2a}{Z}\left[1+\frac{1}{2}\left\{1-\frac{l(l+1)}{n^2}\right\}\right].$$

Find the average value of $1/r$ and show that the total energy of a hydrogen atom is just one-half of the average potential energy.

Appendix

§46. The Dirac delta function

In mathematical physics we often encounter functions which have non-zero values in very short intervals. For example, an impulsive force is envisaged as acting for only a very short interval of time. The Dirac delta function, which is used extensively in quantum mechanics and classical applied mathematics, may be thought of as a generalisation of this concept.

If we consider the function

$$\delta_a(x) = \begin{cases} \dfrac{1}{2a}, & |x| < a; \\ 0, & |x| > a, \end{cases} \tag{46.1}$$

then it is readily shown that

$$\int_{-\infty}^{\infty} \delta_a(x)\,dx = 1. \tag{46.2}$$

Also, if $f(x)$ is any function which is integrable in the interval $(-a, a)$ then, by using the mean value theorem of the integral calculus, we see that

$$\int_{-\infty}^{\infty} f(x)\,\delta_a(x)\,dx = \frac{1}{2a}\int_{-a}^{a} f(x)\,dx = f(\theta a), \qquad |\theta| \leqslant 1. \tag{46.3}$$

We now define

$$\delta(x) = \lim_{a\to 0} \delta_a(x). \tag{46.4}$$

Letting a tend to zero in equations (46.1) and (46.2) we see that $\delta(x)$ satisfies the relations

$$\delta(x) = 0, \quad \text{if} \quad x \neq 0, \tag{46.5}$$

$$\int_{-\infty}^{\infty} \delta(x)\,dx = 1. \tag{46.6}$$

The 'function' $\delta(x)$, defined by equations (46.5) and (46.6), is known as the Dirac delta function. It is unlike the functions we normally encounter in mathematics; the latter are defined to have a definite value (or values) at each point of a certain domain. For this reason Dirac has called the delta function an 'improper function' and has emphasised that it may be used in mathematical analysis only when no inconsistency can possibly arise from its use. The delta function could be dispensed with entirely by using a limiting procedure involving ordinary functions of the kind $\delta_a(x)$, but the 'function' $\delta(x)$ and its 'derivatives' play such a useful role in the formulation and solution of boundary value problems in classical mathematical physics as well as in quantum mechanics that it is important to derive the formal properties of the Dirac delta function. It should be emphasised, however, that these properties are purely formal.

First of all it should be observed that the precise variation of $\delta(x)$ in the neighbourhood of the origin is not important provided that its oscillations, if it has any, are not too violent. For instance, the function

$$\delta(x) = \lim_{n\to\infty} \frac{\sin(2\pi n x)}{\pi x}$$

satisfies the equations (46.5) and (46.6) and has the same formal properties as the function defined by equation (46.4).

If we let a tend to zero in equation (46.3) we obtain the relation

$$\int_{-\infty}^{\infty} f(x)\delta(x)\,\mathrm{d}x = f(0), \tag{46.7}$$

which a simple change of variable transforms to

$$\int_{-\infty}^{\infty} f(x)\delta(x-a)\,\mathrm{d}x = f(a). \tag{46.8}$$

In other words the operation of multiplying $f(x)$ by $\delta(x-a)$ and integrating over all x is merely equivalent to substituting a for x in the original function. Symbolically we may write

$$f(x)\delta(x-a) = f(a)\delta(x-a), \tag{46.9}$$

if we remember that this equation has a meaning only in the sense that its two sides give equivalent results when used as factors in an integrand. As a special case we have

$$x\delta(x) = 0. \tag{46.10}$$

In a similar way we can prove the relations

$$\delta(-x)=\delta(x), \tag{46.11}$$

$$\delta(ax)=\frac{1}{a}\,\delta(x), \qquad a>0, \tag{46.12}$$

$$\delta(a^2-x^2)=\frac{1}{2a}\{\delta(x-a)+\delta(x+a)\}, \qquad a>0. \tag{46.13}$$

Let us now consider the interpretation we must put upon the 'derivatives' of $\delta(x)$. If we assume that $\delta'(x)$ exists and that both it and $\delta(x)$ can be regarded as ordinary functions in the rule for integrating by parts we see that

$$\int_{-\infty}^{\infty} f(x)\delta'(x)\,dx=[f(x)\delta(x)]_{-\infty}^{\infty}-\int_{-\infty}^{\infty} f'(x)\delta(x)\,dx=-f'(0).$$

Repeating this process we find that

$$\int_{-\infty}^{\infty} f(x)\delta^{(n)}(x)\,dx=(-1)^n f^{(n)}(0). \tag{46.14}$$

Defining the scalar product of two functions by

$$\langle f, g\rangle=\int_{-\infty}^{\infty} f(x)g(x)\,dx,$$

we can rewrite equation (46.7) as

$$\langle f, \delta\rangle=f(0) \tag{46.15}$$

and equation (46.14) as

$$\langle f, \delta^{(n)}\rangle=(-1)^n f^{(n)}(0). \tag{46.16}$$

In certain treatments of the theory of generalised functions these equations are taken to define, respectively, the δ-function and its derivatives.

The statement is often made that the Dirac delta function is the derivative of the Heaviside unit function $H(x)$ defined by the equations

$$H(x)=\begin{cases}1, & \text{if } x>0;\\ 0, & \text{if } x<0;\end{cases}$$

and it is easy to see on geometrical grounds that there are reasons for conjecturing such a relationship. To make it precise we note that if, in the definition† of the Stieltjes integral

$$\int_{-\infty}^{\infty} f(x)\, \mathrm{d}F(x),$$

we take $F(x)$ to be the Heaviside function $H(x)$, we find immediately that, for any integrable function $f(x)$,

$$\int_{-\infty}^{\infty} f(x)\, \mathrm{d}H(x) = f(0). \tag{46.17}$$

Comparing equation (46.17) with equation (46.7), we see the relation between $H(x)$ and $\delta(x)$. It may be seen from these equations that $\delta(x)$ is not a function but a Stieltjes measure, and that the use of the Dirac delta function could be avoided entirely by a systematic use of Stieltjes integration.

†The simplest definition of a Stieltjes integral $\int_b^a f(x)\, \mathrm{d}F(x)$ is as the limit of approximative sums $\sum f(\xi_r)[F(x_r) - F(x_{r-1})]$, where the x_r are the points of subdivision of (a, b) and ξ_r lies in the interval (x_{r-1}, x_r).

References

Abramowitz, M. and **Stegun, I. A.,** *Handbook of Mathematical Functions with Formulas, Graphs and Mathematical Tables,* Dover Press, New York, 1966.

Burkill, J. C., *The Theory of Ordinary Differential Equations,* 3rd edn, Longman, London, 1975.

Copson, E. T., *Functions of a Complex Variable,* Oxford University Press, Oxford, 1935.

Coulson, C. A. and **Boyd, T. J. M.,** *Electricity,* 2nd edn, Longman, London, 1979.

Hobson, E. W., *The Theory of Spherical and Ellipsoidal Harmonics,* Chelsea Pub. Co., New York, 1965.

MacRobert, T. M., *Functions of a Complex Variable,* 4th edn, Macmillan, London, 1954.

Mott, N. F. and **Sneddon, I. N.,** *Wave Mechanics and its Applications,* Oxford University Press, Oxford, 1963.

Nosova, L. N., *Tables of Thomson Functions and their First Derivatives,* Pergamon Press, Oxford, 1961.

Slater, L. J., *Generalized Hypergeometric Functions,* Cambridge University Press, Cambridge, 1966.

Tables of Spherical Bessel Functions, (2 vols.) (Prepared by the Mathematical Tables Project of the National Bureau of Standards), New York, 1947.

Watson, G. N., *A Treatise on the Theory of Bessel Functions,* 2nd edn, Cambridge University Press, Cambridge, 1944.

Wendroff, B., *Theoretical Numerical Analysis,* Academic Press, New York, 1966.

Zhurina, M. I. and **Karmazina, L. N.,** *Tables and Formulae for Spherical Harmonics,* Pergamon Press, Oxford, 1966.

Index